THE RIGHT KIND OF CRAZY

The
RIGHT
KIND of
CRAZY

A TRUE STORY OF
TEAMWORK, LEADERSHIP, and
HIGH-STAKES INNOVATION

Adam Steltzner

with William Patrick

Portfolio / Penguin

PORTFOLIO / PENGUIN
An imprint of Penguin Random House LLC
375 Hudson Street
New York, New York 10014
penguin.com

LIBRARY OF CONGRESS CATALOGING-IN-PUBLICATION DATA
Names: Steltzner, Adam, author.
Title: The right kind of crazy : a true story of teamwork, leadership,
    and high-stakes innovation / Adam Steltzner, with William Patrick.
Description: New York : Portfolio, 2016.
Identifiers: LCCN 2015041381 | ISBN 9781591846925 (hardback)
Subjects: LCSH: Steltzner, Adam. | Curiosity (Spacecraft) | Jet Propulsion
    Laboratory (U.S.) | Aerospace engineers—United States. | Roving vehicles
    (Astronautics) | Space vechicles—Landing—Mars (Planet) | Mars (Planet)—
    Exploration. | BISAC: BUSINESS & ECONOMICS / Leadership. |
    BUSINESS & ECONOMICS / Organizational Behavior. | TECHNOLOGY
    & ENGINEERING / Aeronautics & Astronautics.
Classification: LCC TL799.M3 S74 2016 | DDC 629.46—dc23
LC record available at http://lccn.loc.gov/2015041381

Printed in the United States of America
10 9 8 7 6 5 4 3 2 1

Set in Mercury Text G1 with Gotham
Designed by Daniel Lagin

I wish to dedicate this book to Dr. Stephen Prata, my first physics instructor at College of Marin, in Kentfield, California. Dr. Prata shared with his class the thrill and glee he found in the act of discovery. He showed his students that there was a grand, big world out there for us to understand if we followed our curiosity. Curiosity is a spark, and thinking, exploration, and learning are the fires that burn from it. Dr. Prata helped me find that spark of curiosity that led to a fire of exploration and learning that changed my life, burned across the span of time, and put me where I am today, writing of this fantastic journey.

CONTENTS

CONTENTS

THE RIGHT KIND OF CRAZY

Chapter 1

SINATRA, AND A LOT OF BRASS

I'M AT THE JET PROPULSION LABORATORY IN PASADENA, California, in the Cruise Mission Support Area, "mission control," hooking up my iPhone to the Voice Operational Communications Assembly (VOCA), getting ready to launch Ol' Blue Eyes singing Nelson Riddle's 1966 arrangement of "All or Nothing at All."

It's just before 8 p.m. on August 5, 2012, and we're nearing the climax of a massive team effort to land a rover the size of a MINI Cooper on Mars. The overall project, called *Mars Science Laboratory,* has taken ten years to reach this point. Before we're done tonight, we'll have set that rover down gently at our selected site, or we'll have made a smoking crater on the surface of Mars. Whichever way it goes, we won't even know for seven minutes. That's how long it takes data from Mars to reach us back on Earth.

Hunched over the desktop beside me, jamming his VOCA headset into the small iPhone speakers, is my good friend Miguel San

Martin, who is also my deputy in managing our part of the project: the critical component known as Entry, Descent, and Landing (EDL).

The assignment we took on nearly a decade ago sounded straightforward enough: Design a way to deliver a 5,359-pound entry vehicle, which carries a 1,982-pound rover called *Curiosity*, into the Martian atmosphere without harm, then slow it down, guide it to the landing site, and put it down on the surface, safe and sound.

In just a couple of hours—10:32 p.m. Pacific time—we'll know how well we did. It's our all-or-nothing moment, the moment of truth.

As the engineer leading the EDL team, my main job tonight is to be on hand to accept praise or scorn on behalf of the team. Ours is perhaps the most treacherous part of this incredibly complex mission, and definitely the most visible. Like a flaming guitar solo in a stadium concert, EDL might not be the mission's most fundamental element but it is the part that everyone will remember, good or bad, the next day.

Thousands of people have spent a big hunk of their professional lives working fifty- to eighty-hour weeks on this mission; I have fifty colleagues in EDL alone. This is a complete team effort, and an effort of a truly remarkable team at that. But as my friend, mentor, and senior sage at the Jet Propulsion Laboratory (JPL), Gentry Lee, says, there has to be "one ass to kick," and that ass is mine.

At JPL there's a long tradition of playing a "wake up" song at the start of every new day on Mars. Most of the choices have been bright and chipper, like "Don't Worry, Be Happy." But Mig and I are in charge tonight, and the music's our call. As is our custom, we've tried to make a selection that speaks the truth of the situation we're in.

We flip the switch, and our little broadcast goes out through JPL's VOCA system to headsets all over the world. First comes some smooth and mellow brass—trombones, I guess—then a tease of muted trumpets, then more trombones, then another tease from the Hammond B-3, all riding atop that swinging drumbeat. And then the Voice with the Message: "All . . . or nothing at all . . ."

I take a moment to relish the sound, which is totally cinematic. You can see the helicopter shot over Mulholland Drive at sunset with the LA skyline below. Or maybe a '57 T-bird pulling into the Sands in Vegas with Deano at the wheel and Sammy Davis Jr. riding shotgun. It's a skinny-tie, dry-martini arrangement, perfect for a caper like *Ocean's Eleven*—the original, not the remake.

Listening alongside us in mission control are brass of another kind. The president of the California Institute of Technology is in the room, as is the head of JPL (which is an arm of Caltech), along with the head of NASA. Even the film director James Cameron is on hand to see what a real Mars landing looks like.

Since the space shuttle program was canceled in 2010, Mars has been the only game in town for NASA brass, and for space geeks this landing is the World Cup. A couple of large rooms manned by caterers and equipped with large fishbowl windows looking into the control room are filled with an assortment of governors and congressmen and other people wearing suits with flag lapel pins. At one point it looked as if the First Lady would attend, but something must have come up.

Our mission managers selected this date for the landing more than a year ago, based on celestial mechanics and the position of existing Mars orbiters that could photograph our descent and otherwise relay data. The rendezvous was set in motion last November, when the capsule carrying *Curiosity* was launched from the Kennedy Space Center aboard a $300 million Atlas V-51 rocket. This means that our spacecraft—the product of almost a decade of many of our lives and the result of more than ten thousand human years of effort—has been hurtling through interplanetary space toward the Red Planet at about 13,000 miles an hour for the past eight and a half months.

Getting a payload to Mars is hard. The target is anywhere from 40 million to 350 million miles away, depending on the alignment of the planets, and it's orbiting the sun at 24 kilometers a second, more than 50,000 miles per hour. Landing a rover full of delicate scientific instruments at a specific site—in this case the Gale Crater, near Mount Sharp, in the northwestern corner of the Aeolis quadrangle—and at precisely the right moment is harder still. Yet this was the objective of the team of engineers I had the privilege to lead.

The idiosyncrasies of Mars's dust-laden carbon dioxide atmosphere don't make the job any easier. The planet has one-third the gravity of Earth. Its atmosphere is so thin that the air offers little resistance to an object hurtling through it—but is still thick enough to cause tremendous heating when you show up going 13,000-some-odd miles an hour. You'll need to come equipped with both an outsize parachute to slow you down—one that can open while you're still going nearly twice the speed of sound—and a heat shield to keep you from bursting into flames as you enter.

Once that heat shield gets you in, you need to blow it off with pyrotechnics so your radar can see the surface, and then you need the usual rocket thrusters to slow you further and guide your descent. We have been to Mars before, and we have always used this approach. But here's the bummer: The long-legged landers we used for the *Viking* missions to Mars in the '70s could travel safely only to very flat places. We wanted *Curiosity* to follow the scientists' interests, which might take it across the slopes and boulders and surfaces that are endemic to Mars. And the air bags we'd used like industrial-strength Bubble Wrap to drop smaller payloads onto the surface on more recent missions, like *Mars Pathfinder* and Mars Exploration Rover, would only rip, deflate, and smash onto the ground under the weight of a spacecraft this massive.

The solution we came up with: the Sky Crane, an approach that resembled something Wile E. Coyote might rig up with ACME Company products. It certainly didn't inspire confidence on paper. On first hearing about it, Mike Griffin, then the top boss at NASA, memorably said, "I think this is crazy." We convinced him that it might just be the right kind of crazy, but we knew we were taking a risk.

Here's how we needed it to work:

After the nearly nine-month journey of 352 million miles, our work has barely begun. First we have to convert the spacecraft from an interplanetary probe to an atmosphere-tolerating "aircraft." We switch the lander's electrical power from solar to nuclear. The spacecraft must be aligned at the proper angle to withstand 15 g's in deceleration forces and temperatures of 3,800 degrees Fahrenheit when it smacks into the atmosphere of Mars. About 7 miles from the surface, friction with the atmosphere will have slowed the lander's speed from 13,000 miles per hour to roughly 1,000. That's when we pop the supersonic parachute. Twenty-four seconds later, we blow off the heat shield so the radar can see. A mile above the surface, we let go of our parachute and light our rockets to navigate to an altitude of 60 feet. Then—and here's the good part—comes the Sky Crane maneuver, in which the *Curiosity* rover is lowered out of its "rocket backpack" by a set of cables. The two objects, rover and backpack, separated by 25 feet of cable, then descend to the surface. We have to retract the cables right after touchdown, in real time, so they'll stay taut as the rocket backpack continues to descend at a little less than 2 miles an hour. At this point, small guillotines cut the cables, and our rocket backpack, its job complete, flies a safe distance away and crashes into the surface, leaving the rover all alone and (we hope) ready to roll.

If any part of this scheme goes wrong—and all it takes is one failure from among tens of thousands of components to cause catastrophic loss—we will all look like idiots, and I'll be at the head of the line.

So you can imagine that the entire EDL team was already pretty amped up and more than a little bit anxious well before we discovered the glitch.

It was Miguel San Martin who found it. In addition to helping me manage and lead the EDL team at a systems level, Mig served as chief engineer for Guidance, Navigation, and Control. About seventy-two hours ahead of entry, he discovered an error in our "center of navigation," a set of parameters meant to represent the dynamic heart of the spacecraft. The center of navigation is the point from which we make all measurements of dynamic motion. Our onboard computers conduct hundreds of thousands of calculations to determine just how fast the spacecraft is going and in exactly which direction, and they're all based on that agreed-upon starting point.

At JPL we test all our software over a variety of platforms, called test beds. They vary in level of sophistication, the most complete being a copy of *Curiosity*, unglamorously called VSTB (vehicle system test bed), that lives in what we call the Mars Yard, a tennis court–size area of Mars-like rocks and rubble on the JPL campus. But so many parts of the mission and different software developers and teams were constantly struggling for time on her that we made a couple of other versions, not quite as complete, named MSTB (mission system test bed) and FSTS (flight software test set). As we prepared for landing, we ran lots of tests in each of these venues. These test results had to be reconciled with one another and differences understood. As we cruised toward Mars, we also had the actual spacecraft, and we could look at measurements taken on the spacecraft and compare them with data generated within our test beds. They should have all lined up; they should have all matched. Some of them didn't. Mig noticed a tiny discrepancy in the measurements taken on the spacecraft from the value he expected. Which is not how it ought to be.

This difference was reflected in the spinning centripetal acceleration—the acceleration that pulls you toward the outside of a merry-go-round. On the way to Mars our spacecraft spins like a top for stability. The measurements of that spinning acceleration were off by a tiny bit, around 150 micro-g's, or less than the 0.00015 of Earth's gravitational acceleration. It was the kind of discrepancy that another engineer might have blown off completely, but it nagged the hell out of Mig. In 1997 he'd been in charge of navigation and control on *Mars Pathfinder,* the mission that revitalized JPL and launched a new era in Mars exploration. After the spacecraft had landed safely, Mig discovered a time-tagging error in the data coming in from the radar. The discrepancy had not been large enough to endanger the mission, but it had been too close for comfort, and fifteen years later it still bothered him.

Mig's worries had a way of becoming my worries.

When you're leading even one component of a mission like this, you find yourself outwardly defending the reasonableness of your actions while, inside, you're criticizing the shit out of everything that's going down in order to find the one thing that can kill you.

Even though I'd spent nine years of my life on this project, I'd always had a hard time imagining that it could work. The team had spent too many hours trying to anticipate all the ways it would *not* work. My thoughts in the months leading up to the landing were something like this: We are going ahead with this, and I can't imagine that it's going to work, and yet I can't think of a reason that it won't work, and for all I know it will work, but I've also seen it *not* work in other missions, and I know that we don't know everything about the spacecraft because we can't, because it is bigger than all of us.

When a space mission goes wrong, it is rarely the gut-wrenching disaster of *Challenger,* the shuttle that blew up on live television in 1986, killing all seven crew members on board. Everything we know about a mission's success or failure comes from radio signals sent from millions of miles away. If your telemetry goes dead, it might just be a temporary communication error. In the past we've lost contact with a space vehicle and had it come back. But we've also lost contact with space vehicles and never heard from them again, which is the most likely outcome if our craft goes silent during EDL. All we need to declare failure is a persistent lack of data confirming success.

Through most of the journey to Mars, the spacecraft's location is at best an estimate made by radio telescopes on Earth. As it draws closer to Mars, though, the planet's gravitational pull begins to remove uncertainty. We know where Mars is, we know the law of gravity, and this knowledge improves our understanding of where the spacecraft is, but only really toward the last minute (or the last few hours actually). If you've found something that doesn't add up, you have one last chance to make a change. It was in this time frame, this last getting-down-to-the-wire time frame, but completely independent of the Mars approach process, that Miguel found the problem.

After doggedly digging into the discrepancy he found between the spacecraft measurements and predictions, he got down to the essence of the problem, which consisted of three numbers representing the three axes that located the center of navigation. It turned out that when our supplier, Honeywell, delivered the inertial measurement unit—the heart of the guidance system—a JPL guy made a mistake logging in those three numbers. Rocket science is a high-tech world, but it's run by people, and people make mistakes.

During our spacecraft's long flight, we had regularly scheduled meetings to discuss software parameters we might want to tweak as we approached Mars, including trajectory parameters. Dust storms on Mars are a nightmare, and at all times we have a roomful of people obsessing over atmospheric updates, not just for the landing site but for the entire planet, in case we need to fine-tune the flight path we come in on. These types of parameters we had planned on changing, and we had structured the software to make it easy and safe to do so.

Parameters more at the core of our software, like the center of navigation numbers Mig was looking at, were not in that set. We could change them, sure, but it was a risky proposition. When you're a couple of days from the end of a nearly nine-month voyage, following nine years of development, you're not eager to tamper with anything unless there's a damn good reason. You certainly don't blithely re-jigger vital software parameters, because tinkering with something as simple as the date or time might inadvertently alter one of the thirty thousand other parameters and cause a catastrophe.

But now Mars is looming large in our windshield. We've made it this far, and we've landed successfully in all the simulations that contained the error, so does that mean we should live with that error? Should we alter the parameters, or should we let it ride?

Mig found the error Wednesday and had confirmed it by late Thursday night, August 2. We immediately set up a "tiger team" of about twenty-five specialists to drop everything and launch a full investigation of the anomaly response—spacecraft ops–speak for "Look into this and get it un-fucked-up if possible." Folks from Guidance, Navigation, and Control, from software, and from trajectory simulations broke the problem into pieces so that subteams could pursue multiple lines of attack, and the subteams began pulling all-nighters. We ran trajectory simulations using both the correct parameters and the erroneous ones. When we compared the results, we couldn't find any differences. That didn't mean they weren't there. It just meant they didn't show up in our simulations. The software folks spent their efforts making sure we could fix the parameter error without upsetting the rest of the flight software. Everyone crammed furiously until 5 a.m. on Saturday, August 4, when we assembled, pencils down, for a come-to-Jesus session. Two hours later there would be a second meeting, with the project managers, to make the actual decision. Do we correct the error or not?

Navigation and control is not a trivial matter, and our center of navigation was off by about three inches. Was that enough of a discrepancy to slam us into the Martian atmosphere at the wrong angle and burn up the spacecraft? Was it enough to cause us to miss our landing site and put us down in mountainous terrain where we might smash into a mountain or topple over? Years of man-hours and billions of dollars were at stake.

Our Saturday-morning "What do we think?" discussion ran long. While everyone found the error unsettling, no one argued for taking the risk to correct it. It could mean nothing in the end, or it could poison us in a way that we could not imagine. But no one had isolated the one glaring truth that said this thing was going to kill us unless we fixed it.

We were still going around and around on the engineering analysis at 7 a.m. when the seven senior managers, all looking fresh and rested, showed up. The top dog for the mission was Pete Theisinger, silver haired and slight of build but a tough fighter. I told him I hadn't polled my team yet. "I'm happy to do it right now, in real time," I added.

Pete agreed, and so, with the senior decision makers looking on, we went around the table to get everyone's best judgment. The poll-taking would end up with me.

Everyone said, "Steady as she goes." Oddly enough, this included Miguel, who'd first spotted the problem and had obsessed about it until he got to the source.

Pete looked very relieved with our group's seeming endorsement of the status quo. He began to move the meeting toward closure. But then I stopped him.

"I said we'd poll the team," I said. "Unfortunately, the team's split . . . because I think we should make the change."

I could see the anxiety tighten his face, and then he settled back to consider his options. The whole EDL team has said that we're okay as is except for me, the guy in charge of that team, the guy specifically tasked with landing the spacecraft. I wasn't an unassailable authority, but I was the one ass to kick.

Pete then polled his colleagues—the rest of the senior management team, including engineering leadership. One after another they echoed the tiger team, somewhat sheepishly admitting that they would rather not make the change. This reluctant chorus of "steady as she goes" continued all around the room until we got to the last two guys: Richard Cook, the deputy project manager, and Rob Manning, the project chief engineer.

Richard said, "I agree with Adam. I think we have to make the change."

Then Pete turned his eyes to Rob. "What say you, chief engineer?"

"I'm with Adam and Richard. We've practiced making these kinds of changes. Let's do what we've practiced and make this right."

Pete drew in a long breath, as if absorbing all the data that had been crunched in the past thirty-six hours, along with all the opinions expressed, leavened with fifty years of life experience. Then he said, "We'll make the change. Prepare the command for transmission, and radiate the command."

So just before 9 a.m. Saturday, thirty-nine hours after we'd begun our assessment, we altered the three numbers that pinpoint the center of navigation. Fourteen minutes later—the time required for a round-trip transmission between Earth and near Mars—the spacecraft reported back that the change had been received and the update completed. At least as near as it could tell.

Had we just screwed the pooch, or had we averted an "O-ring moment" (the culprit behind the *Challenger* disaster) and saved the entire mission? There would be no way to know for thirty-six hours. During that time the spacecraft would have traveled 400,000 more miles and gone through the torturous seven minutes it would take for the rover to enter Mars's atmosphere, descend to the surface, and land on the Red Planet safely—the Seven Minutes of Terror.

———

Spoiler alert: We made it. When the seven minutes came and went and we heard that first ping back from *Curiosity,* we looked like geniuses and everybody loved us. (You can skip ahead to the last chapter if you want more detail on who cried, who cheered, and how we stormed the press conference.) Whether the last-minute change made the difference between a success and history's most expensive pile of burning scrap metal, we may never know.

What we do know is that a roomful of the world's best aerospace engineers couldn't agree on whether it would make a difference. Engineers are supposed to be rational minimalists. Did each of us do a purely rational calculation and simply come up with different answers? Did we use our slide rules incorrectly? Were our calculators broken? Was it really just a question of math?

I don't think so.

Back in the early days of the American space program, the engineers on TV during the rocket launches and moon landings all sported white short-sleeved shirts and skinny ties with tie clips and nerd eyeglasses, all of which contributed to an "I am a robot" image that might have been meant to intimidate the Soviets. But engineers aren't robots. Engineers are people, and no matter how carefully we pursue the truth of the physical universe as reflected in the hard data, our interpretation of the physical laws as applied to the business at hand is always going to be filtered through who we are as individuals. Any truth we arrive at, no matter how diligently we pursue it, is always going to be an approximation, a model of the universe and not the universe itself. When problems get thorny and the outcome or correct path is uncertain, as it was for the group contemplating changing software parameters at the last minute, human judgment and an appreciation of the limits of your understanding become paramount. This is true whether you are engineers building a spacecraft or software developers working on the next big iPhone app.

Beyond the importance of human judgment is the need to understand what makes other humans tick. The English word *engineer* comes from the Old French *engin,* meaning "skill or cleverness." The act of engineering is to solve a problem by being clever and utilizing our understanding of the world around us. In an era of projects with billion-dollar budgets and head counts in the thousands, part of "the world around us" that needs to be understood is the people involved. If you're going to succeed in that kind of creative, collaborative environment, especially if you're going to lead and manage the development of something innovative, you need to engage a lot more of yourself than your knowledge of fluid dynamics or stress analysis.

As a practical enterprise, engineering is hugely dependent on honestly facing the hard data. If you pursue your own personal "truth," or if you settle for a partial or parochial truth, or if you deny the truth because it's awkward or inconvenient, your bridge falls down. Your cyclotron doesn't find any particles. You never get your spacecraft anywhere near Mars, much less safely on the surface.

But it's a mistake to assume—which plenty of folks, including engineers, often do—that the answers live in some preexisting space, that all you have to do is put the right equations on the table, that you don't have to use judgment or anything associated with the emotions to get to the solution. The best problems are simply too complicated to have a clean equation that describes them. In the real world of budgets and politics and the dynamics of large organizations, getting anything done is a no-holds-barred brawl, and as in any street fight, success or failure is a function of everything you bring to it: cleverness and intelligence, knowledge and technical prowess, charisma or awkwardness, shyness or the ability to persuade, self-confidence or self-doubt, self-awareness or denial.

On its face, this is the story of an audacious engineering project: the design and construction of a hugely complex rover and the innovative, "crazy" landing system that delivered that rover to Mars. But that's just the basic plot.

This is also a personal story about how I ended up at the Jet Propulsion Laboratory building spacecraft and how I learned from my time at the lab to lead an extremely talented team to solve impossible problems. It is a story about harnessing human curiosity to build something truly fantastic and about being honest enough about human nature to protect ourselves from self-deception on a scale that could bring disaster. And it's an exploration of the thought processes, leadership techniques, and problem-solving skills that went into making such an exceptional effort possible.

To the extent that my perspective might provide some insights, it is also going to be flawed. I'm going to present the story as honestly as I can, but my version is always going to be limited—colored by the lens of my perception.

Most, if not all, of the great works of our species have been team efforts. If we want to do great things—whether it's slow global warming, end malaria, or put a human on Mars—we can't rely on the lone genius working out of his garage. We need to figure out how to engage people of diverse talents, perspectives, and worldviews to come together to produce great work. My ultimate goal for this book is to provide a fresh perspective on how leaders can successfully engage smart people to build challenging, high-stakes, innovative projects. It's my hope that these observations and lessons are transferable to others' efforts in other fields.

Beyond that, I hope you will find a reflection of your own humanity in the story of the work that we do to explore our universe—the work of extending the edge of what we know and perhaps even who we are as a species.

Chapter 2

HOW CURIOSITY CHANGED MY LIFE

I SOMETIMES JOKE THAT CURIOSITY CHANGED MY LIFE. BUT IN some sense it is really true. I am very different from the man I could have been, and I think I might be better. I'm sharing this story not because I think it's unique but because I think it's not. There are scores of young people out there getting bad or conflicting advice, locked into what they think is the thing they are supposed to do and pushing themselves to perform. With those high standards frequently comes a crippling fear of failure. I was crippled by it, and if I hadn't been curious about the world, I would probably still be working in a health food store, dreaming of the day when my band would hit it big.

Yes, I have helped land robots on Mars, but I was not one of the math geniuses who carried a briefcase and won physics prizes in high school. I knew some of those guys, I liked them, but I was not part of their crowd. Some may consider it a miracle that I ever got my diploma. My only good subject was theater, and instead of studying, I smoked pot, rode mountain bikes, chased girls, and played gigs with my band. I had no real ambitions except perhaps to magically become Elvis Costello or Joe Jackson.

I grew up in the sixties and seventies, mostly in Sausalito, just north of San Francisco, with parents who were very much in tune with the time and the place. My mother was a true free spirit. She lied about her age to serve in the Women's Army Auxiliary Corps during World War II and afterward was a bit of a beatnik in San Francisco, managing nightclubs and dating influential female artists of the time. Her intelligence and independence aside, she still preferred to sit in the passenger seat when there was a man to drive. She met my father when, fresh from Army service, he stopped for a drink at the bar of a hotel she managed in the foothills of eastern California's Sierra Nevada. Sparks flew, and he stayed four days, leaving my mother with the address of an accountant in Piedmont to whom the bill should be sent.

My parents, my brother, and I lived an almost comfortable existence on what was left of my father's family inheritance. My father was a great reader, knowledgeable on subjects ranging from the aerodynamics of laminar airfoils to the artificial insemination of sheep, but he never showed much interest in earning a living. Work meant struggle, struggle meant the possibility of failure, and failure was simply unacceptable.

As the family wealth dwindled, denial hung over us all like a cloud. My father dealt with it by drinking heavily. I coped by becoming a human projectile.

At a very early age, I pursued every form of physical recklessness, from skateboarding in traffic to biking down steep mountain roads at race-car speeds. I was the kid who climbed every fence and leapt off every ledge. In retrospect, I think I was trying to escape the fear that seemed to cage my father, or to suspend it via an act of will and perhaps grab on to something real and true. But mostly I broke bones—thirty-two in all before I was grown.

The closest thing to a hint of a future career in space exploration was my daydreams, which dealt with spatial reasoning. My number one alternate reality was a weird game in which gravity would turn off, then turn on in a different direction. I would be left trying to understand this new force field and the trajectories that objects, especially myself, would now take through space. Sometimes gravity ran sideways and, in my imagination, would slam me into a corner of the room. I would have to grab a windowsill or try to avoid a column.

Meanwhile, in my waking life, there was no such thing as following the customary path from point A to point B in the neighborhood. One game was to follow a straight line no matter how many obstacles were in my way, refusing to bend my will to the designs and will of "the man." Did I have to climb three 8-foot-high fences and make my way through a couple of industrial yards on my way to get a candy bar? So much the better. Climb along the 25-foot-high peak of a roofline? No problem. Property owners would put axle grease on their fences and catwalks to keep me off, but that just increased the appeal. Let's see if I can now jump off that roof to that gate and walk along the top of it to the fence without touching the grease. "Get off the roof, Adam!" was a common cry from the neighbors. And "Hello again, Adam," a common greeting in the ER.

By the time I reached high school, it had become abundantly clear that I was not "college material." My parents sent me to therapy to explore my combination of extreme risk taking and minimal consciousness in school, but I don't think my behavior was really all that great a mystery. I was simply acting out a challenge to my dad.

I could see what fear had done to him, and I was having none of it. And if he didn't have to go to a job and work hard, why should I? I'd also internalized the message from my dad that trying to do anything seriously was pointless, even dangerous, because absolute perfection was required, and absolute perfection was unattainable.

After I barely managed to graduate from high school, I moved to Mill Valley, took a day job at a health food store, and started playing bass in a kind of rock-jazz-punk fusion band called Exit.

One night I had a gig at a club in Corte Madera, on the way up to Larkspur, so early in the evening I was driving up Highway 101, heading for our sound check, and I noticed the constellation Orion over my right shoulder, meaning that it was in the eastern sky, looking toward Point Richmond. After midnight, when I was driving home, there it was again over my right shoulder, meaning that now it was in the western sky, looking toward Stinson Beach. I'd never taken an astronomy course, but for some reason this intrigued me. We see the stars "moving," but I knew that what's actually happening is that *we're* moving, riding on the surface of the Earth as it rotates on its axis and orbits the sun—or did I? Was that what was really happening? How does all that shit really work?

As I drove on home that night, I kept thinking about this movement in the sky—what I'd now call celestial mechanics—and for the first time in a long, long while, I was deeply curious about something. And that curiosity changed my life.

A couple of days later, I decided to drop by College of Marin, the local community college, to see about signing up for an astronomy class. Coincidentally, it was the day of the next semester's course registration, and sign-up sheets covered the walls of the gym. I grazed the offerings until I saw what I was looking for: Introduction to Astronomy. But then I read the fine print. To take the course, you had to have taken the prerequisite: Physics 10. I signed up for both, hoping we could work out some sort of deal. As it turned out, the joke was on me, because not enough people signed up for the astronomy course, so it was canceled, and there I was in a physics course I'd had no intention of taking. I decided to stick it out because, deep down, I craved a new experience. That and the fact that one of my classmates was a high school buddy I used to race mountain bikes with.

I didn't have a particular game plan. I was simply following my curiosity. I had let go of any expectation about where my efforts might take me. I certainly had no expectation of great success. I wasn't waiting to become Elvis Costello anymore. I was simply doing my best to learn about my world and the universe around me. In some sense I was starting small, starting at the ground floor with the bedrock truth of the physical universe and hoping to build up from there.

I was also experiencing what my friend Miguel calls a "constructive interference of personality disorders," which he claims is the secret of success for most people who get anywhere. But be that as it may, at College of Marin math and physics opened up to me a shiny new world of precision and clarity and absolute honesty. The idea that there were laws that governed this confusing universe we inhabit was a huge turn-on.

I'd taken elementary algebra and geometry (twice) in high school, passing with an F+. I'd done whatever the minimum requirement was to graduate, but anything I'd learned had gone in one ear and out the other, leaving no trace. Luckily, Physics 10 was "physics for poets," which meant there was basically no math, so I thought that maybe I could survive.

The larger stroke of luck was that the man teaching the course, Stephen Prata, had the gift of being able to share with his students his passion for understanding the universe. When Dr. Prata took a piece of chalk and scrawled F = MA on the blackboard, it seemed to me like an incantation, making mass (M) and acceleration (A) transmute and reveal their nature. I'd experienced the force (F) he was talking about riding skateboards, so I understood in my gut what the equation was saying. Now I could see how that fact of nature could be analyzed abstractly. You could make a model, and then you could use that model to make predictions. And then you could use those predictions to make stuff or to make stuff happen. You might even use them to explore outer space.

To me this manipulation of equations that represented the functions of the universe was not only magic but a thing of intense beauty. It seemed to speak the hard truth that I had been looking for when I was slamming myself against the world and breaking all those bones. But I was so far behind in my education that I would have a long, long way to go before I could approach the truth in the form of an equation.

I worked hard to catch up, entering a monkish phase during which I let go of music and buzzed off all my hair. I kept my apartment and my day job at Living Foods in Mill Valley, and I rode my mountain bike back and forth over Mount Tamalpais to attend class. It was what a Jungian shrink might call a time of ashes, when you go down to get real in order to go back up. But I never really saw it that way. I was simply saving my life and the lives of everyone to follow in my lineage, hell-bent on not repeating the mistakes of my father.

I did well at Marin—the feeling that you're playing out a family curse if you don't stick with something is excellent motivation—and it awakened a dormant competitiveness within me. I started working harder to see just how well I could do.

After a few years, I transferred to the University of California, Davis, where I majored in mechanical engineering and design. The way I looked at it, being an engineer was like being a physicist, only you could count on getting paid. But what appealed to me most was the focus on fearlessly seeking out the truth of a situation, being objective and empirical. It would take a while to realize that there was a lot more to getting engineering done than a firm grasp of objective truth.

Three years later I graduated and gave the valedictory address for the College of Engineering. I was offered a full ride for graduate school from all the right places—Stanford, Caltech, MIT—but when I visited Caltech, in Pasadena, they seemed focused on the work more than on their prestige, and it scared the crap out of me. These guys were *serious*! I chose Caltech. For someone with decent math skills and something to prove, entering the California Institute of Technology was like joining the Navy SEALs.

In fact, I had no burning desire to do research, and I didn't even have a particular field that I wanted to explore. I was driven simply by curiosity and the need to overcome my insecurities.

I never found a professor I clicked with or research that excited me, and with a vague dissatisfaction bubbling up, I went to see a former student give a seminar on his work at Orbital Sciences Corporation in Maryland. They'd developed the Pegasus launch vehicle, designed to carry small satellites into space and also designed to be launched from an L-1011 airliner at 30,000 feet or so to bypass the denser parts of Earth's atmosphere.

My work at Caltech up to that point had been theoretical and abstract. Until I saw the talk from the guy from Orbital, I hadn't

realized how hungry I was for the hands-on application of engineering training. I wanted to get my hands dirty. I wanted to build
something. By the end of the summer, I'd made up my mind to
move on.

I left Caltech with a master's in applied mechanics as a kind of
consolation prize for one year of work toward a Ph.D. Then I was
faced with the question of what to do with myself.

For the past few months, I'd been dating a fellow graduate student named Ruthann. She was staying in school, and I wanted to be
near her, so I needed an engineering job in the LA area, which at the
time, the early nineties, meant defense contracting. But I didn't
want to build weapons systems.

The next logical choice was the Jet Propulsion Laboratory. Nes-
tled on 177 acres rising up into the San Gabriel Mountains, JPL was
just a couple of miles on the other side of the Foothill Freeway from
Caltech. Federally funded but actually managed by Caltech, JPL
had five thousand employees and, at any given time, maybe a thousand additional contractors.

As an undergrad I'd read a little science fiction, including Carl
Sagan's *Contact,* in which the young heroine works at a government
lab and flies around trying to decipher what appears to be a message
from aliens trying to reach us. I figured working at JPL might be a
little like that.

The reality proved to be not quite "ET phone home," but the lab
did have an interesting mystical-cosmological vibe that dated back
to its founding fathers.

In 1926 Caltech was one of the first universities to establish an
aeronautics lab. There, a grad student named Frank Malina pio-
neered an alcohol-fueled rocket motor, which at the time must have
seemed like something out of Buck Rogers.

Malina's experiments were considered too dangerous for the main campus, so he and a few other rocket enthusiasts, known as the Suicide Squad, moved their operations a few miles away to a dry creek bed where an explosion or two wouldn't matter much. Eventually, that isolated spot, the Arroyo Seco, became the site of the JPL, with Malina as its first director.

The act of engineering, of being ingenious, is fundamentally a creative one. Although they're not frequently thought of in the same stroke, I think engineers and artists have a great deal in common. Malina, who would later become a sculptor, was the perfect example. Another was his Suicide Squad mate and fellow JPL founder Jack Parsons, a self-taught chemical engineer and a onetime Marxist who converted to a religious movement called Thelema, embodying the freethinking, boundary-pushing craziness of that era.

Before each rocket blast, Parsons would invoke the "Hymn to Pan" ("Thrill with lissome lust of the light, O man!") written by the occult poet and Thelema founder Aleister Crowley. Over time Parsons became leader of Ordo Templi Orientis (Order of the Temple of the East), the California branch of the religion, which brought him in contact with Scientology founder L. Ron Hubbard. Together they championed a new era of free love, which, to their minds, was the natural consequence of breaking free from the shackles of four-dimensional space-time. Joining their effects was an "elemental mate," in this case a redhead named Marjorie Cameron, who would later star in films, including an occult film by Kenneth Anger.

Along with radical politics and offbeat religion, the early days of JPL included a lot of beer around campfires on the arroyo and drunken brawls over engineering issues. Even inside the office, the culture was open and free spirited, reflecting the freedom inherent in exploring the unknown. In the late 1930s, with World War II coming on, the U.S. government was interested in developing rocket-assisted takeoff for fixed-wing aircraft, and it began funding the research begun by Malina, Parsons, and others. This led to contracts with the Army Ordnance Corps to develop the Corporal and Sergeant missiles.

Eventually Malina became disillusioned with weapons research and dabbled further in leftist politics, which led to his being investigated by the FBI, which led to his moving to France, dropping rocketry, and becoming a kinetic sculptor. Parsons's life on the ragged side of crazy finally caught up with him: He lost his security clearance and during the McCarthy era was reduced to doing menial jobs. He died in an explosion at his workshop, and the jury's still out as to whether or not it was an accident.

Through the midfifties, the lab continued to do cutting-edge guidance and propulsion for military missiles. The freewheeling spirit of the place persisted in all-night poker games on the train to the missile range in White Sands, New Mexico, with restorative detours to sample the south-of-the-border pleasures of Juárez.

Even through all the years of military work, JPL remained fiercely independent, and with the esprit de corps and the balls to go ahead and do things other people thought of as crazy. When the race for the moon was on as the Cold War escalated, William Pickering, the lab's director from 1954 to 1976, decided that JPL should devote itself to interplanetary missions—to those "other" objects that no one seemed particularly interested in. In a lot of ways, JPL was like those fifteenth-century explorers who simply sailed over the horizon—not knowing.

The lab designed and built *Explorer 1*, which became the nation's first orbiting satellite. This success brought JPL under the aegis of the newly formed National Aeronautics and Space Administration and led to the *Ranger* and *Surveyor* projects that prepared the way for *Apollo,* the first manned missions to the moon. This was followed by the *Mariner* missions to Venus, Mars, and Mercury, and ultimately, the *Viking* landing on Mars in 1976, as well as the *Voyager* missions through the outer solar system, which are flying in interstellar space today.

To me, JPL was always a source of mind-blowing images of other worlds. The first images of Jupiter and Saturn, huge images that made you feel like you were right there beside the planet, filled me with awe as a youth. And with JPL's deep currents of outrageousness and iconoclasm, it definitely seemed like my kind of place.

Ruthann's dad worked at JPL, but I didn't want to ask him for help. I wanted this to be my deal, and I knew he would take it too far and try to "fix me up." Instead, I asked my professors for contacts, and one put me in touch with Donald Bickler, a famously curmudgeonly engineer who would become the developer of the six-wheeled-rover architecture that we would later use so successfully on Mars. We talked on the phone, I sent a résumé, and he told me to drop by.

On the day of the interview, back in 1991, I drove up past the U.S. Forest Service Ranger Station, then the stables and corrals of the Flint Ridge Riding Club. This was still dry and dusty "cowboy country," and some of the low-slung buildings at the top of the hill dated back to the fifties. One step up from Quonset huts, they had a sort of Oppenheimer–at–Los Alamos look and feel. The newer, taller ones, built at slightly lower elevations, had the conventional glass and steel of high technology. Bickler's office was in an odd two-story, utilitarian building with nearly no windows.

Bickler was an iconoclast who enjoying taking the intellectual establishment down a peg whenever he could. As I was coming straight from Caltech, maybe in his mind I represented that establishment. For whatever reason, it wasn't long at all before we got into our first disagreement. I think it had to do with the torsional stiffness of a bicycle frame. I knew this stuff firsthand—I'd built and raced bikes as an undergrad—but that didn't impress Bickler. Then he asked me about creating a wheeled robot that would not lose traction. I knew from all my breakneck trips down Mount Tamalpais that a human is doing an incredible number of calculations of the capacity of the ground to support the various amounts of traction, but at super high speeds, of course, you simply lose yourself in the flow. So my suggestion involved putting a human in the loop. Bickler didn't like my answer. By this time I realized the interview was over, and Bickler unceremoniously asked me to leave. "You can find your own way out," he said, and I stumbled through the lab building looking for the exit.

I was floored. So I wasn't good enough to work at JPL?

In the years since, I have worked arm in arm with Bickler without a hitch, and I've realized that the interview was an epic fail on my part. I went in assuming I'd be right, and my cocksure attitude prevented me from really hearing what Bickler was trying to ask. JPL embodies a culture of questioners, and my unwavering certainty in the face of doubt probably didn't communicate that I was ready to give up my pet answers in pursuit of real truth.

Bickler's was the only name I'd been given, so I took out the Caltech/JPL phonebook and looked for sections in the JPL organization that sounded like they did stuff that I'd like to do. I sent my résumé off to these folks through the campus/JPL mail. Then I followed up with a call.

First on my list was a guy named Mike Lou, in the Applied Mechanics Technologies Section. I told him that I had sent him a résumé a week earlier and asked if he would like to interview me for any possible openings. On the other end of the phone, I heard him pause, then I heard the sound of shuffling papers. When he came back on, he said, "I'm sorry, but I have a stack of résumés on my desk I'm interested in, and I don't see yours. So we must have already filtered you out. Sorry. Wish you the best of luck in your search."

I was crushed. Who are these guys? What titans work at JPL that they would pass on the top student from the College of Engineering at UC Davis, who was also in the top 20 percent of his class at Caltech? Isn't that enough? What does it take?!

Five minutes later the phone rang, and it was Mike Lou again. He had found my résumé in his unopened mail, and apparently it looked good enough for him. He wanted to set up an interview. Happily, there were no quizzes on torsional stiffness, and I approached the interview with a huge portion of humility. Which is how my career at the lab began.

So curiosity had led me to a place I could never have imagined as a kid. I wasn't Elvis Costello, and I wasn't ever going to be, but I did have a job at the lab responsible for the bulk of the U.S. unmanned space effort. They were the guys who built all the spacecraft that brought back all those incredible images that had struck me with awe in my youth—Jupiter's red spot, volcanoes on Io! Pretty fucking cool.

Truth is, it wasn't curiosity alone. There was certainly a great deal of hard work and persistence, and there was a great deal of struggle within myself, wondering if I could do it. Under all the angst, though, whenever necessary, I could strip down what I was working on and find that glimmer of curiosity about the question. Curiosity is a spark, and exploration is the fire that burns from it. I was beginning a whole new chapter of my life, a whole new exploration into the unknown.

Chapter 3

HOLD ON TO THE DOUBT

WHEN I TOOK PHYSICS BACK AT COLLEGE OF MARIN, WE were allowed to bring one sheet of paper into each test. On that one sheet, we could have anything we wanted. Most students would take a 000 (ultrafine-point) Rapidograph pen and write down every equation in the book. My sheet of paper might list a few of the governing laws we'd been studying, but most of the page was devoted to the words "Hold on to the doubt!" usually written in bright colors with a highlighter pen.

Through my dad I knew full well that fear of failing was enough to cause me to fail. Staring at the blank space beneath the problem was sufficiently anxiety provoking that it could lead me to rush to an ill-considered answer. Jumping too soon into solution space might cause me to miss some key element of the problem being posed or to answer a conceptually adjacent question.

I resolved that the solution for my anxiety while staring at the unanswered question was to remain calm in the presence of the openness, to not close off the inquiry too soon and thus run at full speed into a solution that might not take the whole truth of the problem into account. Holding on to the doubt meant listening to all that the problem had to say and not making assumptions, and committing to a plan of action based on them, until the deepest truth presented itself.

It was a philosophy that would serve me well in my time at JPL.

When I started working at the lab, in October 1991, it was on the cusp of a seismic shift in culture. In the beginning the lab had been lean and mean, a bastion of independence and personal creativity, but as it grew it became more process-oriented and took on larger, more unwieldy projects. By the early 1990s, there were far more budget meetings than brawls around campfires. A sophisticated administrative infrastructure, often referred to as a "matrix" organization, surrounded the technical work. For an entering twenty-eight-year-old, the matrix was not super inspiring. Your work came to you at an emotional remove, through a long line of folks playing the "telephone" game, and you were not particularly empowered or encouraged to take ownership.

No one at JPL knew it at the time, but we were about to enter a period of change and rebirth. Driving the change was a new NASA administrator, Daniel S. Goldin, the former general manager at aerospace giant TRW's Space and Technology Group. Goldin believed that NASA itself had become slow, expensive, and not necessarily reliable. Projects were overambitious, budgets bloated, bureaucracies massive. As for planetary exploration, we had already had at least a fleeting encounter with just about every destination in the solar system. Besides, any "space race" justification for the lab's work had been overtaken by history. The Soviet Union had collapsed. The Cold War was over.

In 1992 Goldin gave a speech at JPL challenging us to adopt an approach he called "faster, better, cheaper." Instead of pursuing a handful of huge and expensive missions at NASA's behest, JPL would go after a variety of smaller missions. We would also compete for funding with NASA's proprietary labs, like the Ames Research Center near San Jose and the Langley Research Center in Virginia. Goldin wanted smaller teams on everything. Not ten people checking one guy's work, but one guy checking it ten times. "How much process can you throw away?" he asked.

"Faster, better, cheaper" also meant more risk. "A project that's twenty for twenty isn't successful," Goldin said in his speech. "It's proof that we're playing it too safe. If the gain is great, risk is warranted. Failure is okay as long as it's on a project that's pushing the frontiers of technology."

Over the next decade, NASA's budget would shrink by 18 percent and JPL's head count would be reduced, from 5,500, by about 1,000. On the plus side, "faster, better, cheaper" meant a return to the lab's more freewheeling roots, with ownership of the projects driven back down toward the troops and JPL returning to the kind of place where someone like me would want to spend a career.

At the time JPL was just a job for me. As luck would have it, I landed in one of its more conservative bastions, working as an analyst in the Spacecraft Structures and Dynamics Group. Structures and dynamics is the bread and butter of the Southern California aerospace industry, and of the fifteen or so people in my group, about half were hired guns who had made the rounds of the big companies—Northrop, Rockwell, TRW—where Dockers and polo shirts were about as radical as they could get. Overcompensating, perhaps, on my first day at work, I wore black slacks, a white dress shirt, and a blue tie. I stuck out like a sore thumb.

Nobody seemed a likely hero or heroine for a techno-thriller like *Contact,* but everyone had more to his story than you might expect. My new boss, Frank Tillman, was a quiet, older, African American, keep-your-head-down stress analyst who restored old E-Type Jag fastbacks in his garage at home. Faramarz Keyvanfar, another "stress man," a blue-eyed, blond Persian Jew from Tehran, turned my notions of region, religion, and ethnicity on their ear. Darlene Lee had been a grade school teacher until she got bored, went back for a degree in mechanical engineering, and then worked for Boeing up in Seattle. None of these folks fit the cardboard-cutout stereotype of engineer.

I shared a dingy twelve-by-fifteen office with two other engineers, Rob Calvet and Celeste Satter. Rob had done his undergraduate degree at Caltech and was super bright, kind of what you'd expect from a dyed-in-the-wool engineering student. Celeste had pictures of cats on her desk and seemed like she might regularly attend Star Trek conventions. Later I'd find out she was a world champion pistol markswoman.

They were both about ten years older than me and loved working as analysts. I soon learned that being an analyst was essentially a compliance role—hardly anything terribly inspiring or demanding, except in terms of pure technical expertise.

Here's how it worked: Engineers from JPL's designing wing would design a component and throw it over the wall to one of us to check it. They did the creative work; we analyzed the structure to see if the design would hold up.

This work demanded incredible patience. Structural analysis involved sitting quietly and asking the same question over and over again with different boundary conditions. I had a hard time getting used to the routine. With the three desks squeezed together and all the silent tabulations, Rob, Celeste, and I were like clerks in a Victorian countinghouse, only with computers instead of ledger books.

The first work that came my way was to analyze the structure of stratospheric balloon gondolas. Scientists pack these with instruments, frequently telescopes, then lift them above most of Earth's atmosphere, up to 120,000 feet or so, to view distant objects in space. A failure of one of these gondolas had prompted the National Scientific Balloon Facility to rewrite the standards, and it was my job to do a stress analysis of all the hardware to make sure the gondolas we were using could handle the updated loading requirements. I dutifully analyzed the stresses in the structural members—struts, bolts—and calculated how close the material was to breaking.

Despite my zeal for digging down and ferreting out the objective truth, I found that I was not super interested in the kind of engineering where the next problem looks just like the last one. These were problems where, if a hundred people input the data correctly, the output would be the same every time. For that you could write a computer program—and indeed, in the twenty-first century stress analysis has become largely automated.

I worked on gondolas for about six months, and while it wasn't exactly being Elvis Costello, it was a hell of a lot better than working in a health food store in Mill Valley. I was doing something real, not selling the promise of some hippie dietary supplement. This engineering thing was all about making ideas reality. What's more, I started to love the JPL weirdos I worked alongside.

During that first year, Frank assigned me a mentor, a guy named Bob Norton, who had helped develop finite element analysis, a computer technique for calculating stresses. About fifty-five, with short, cropped salt-and-pepper hair, he seemed like the central casting version of the pale, white-guy, "I am a robot" engineer from the fifties.

One day I wore a Django Reinhardt T-shirt to work, and Bob said, "You like jazz?"

He invited me over to see his record collection. We listened to some music, then he took me out to the garage to show me his Sunbeam Tiger. This classic British sports car originated as the Sunbeam Alpine, a little roadster, but then muscle car designer Carroll Shelby got ahold of it, extracted the four-cylinder engine, and replaced it with a honking Ford V-8.

One thing led to another, and later that evening the middle-aged straight arrow and the twenty-eight-year-old punk in shorts and Doc Martens were doing speed trials on a secluded stretch of highway, peeling out and doing zero to sixty in sub-five seconds.

Shortly thereafter Bob recommended that, as part of my training, I spend some time in the field. Out in the Mojave Desert at Daggett Field, near Barstow, we were about to launch a balloon to carry a gondola I had analyzed, which would drift with the winds over the unpopulated areas of Arizona and New Mexico, then be brought down in West Texas. The stated rationale was for me to do a visual inspection of the gondola, but that was total nonsense. Bob told me later that he just thought it would be good for me to get out of the office and see some engineering that was not done at a desk.

For my junket into the desert, I wound up with a massive Lincoln Continental painted NASA white from the JPL motor pool. It had all the plush appointments inside and government plates. I felt so strangely special driving it—and so off the leash. I pointed the car north toward the Mojave, and in my best approximation of Hunter S. Thompson, drove the car 80 miles an hour into the desert. Though we spent two days waiting for the launch, the winds failed to cooperate, and it never happened. But those long hot days, stretched out reading in the backseat of a NASA white Continental, helped me start to feel part of the space game. Images of *The Right Stuff* echoed in the clear blue skies and vast desert vistas. I started to feel part of a multigenerational effort to push the boundaries of our exploration always further, farther, deeper into space.

What Bob and I had done—which was to look beyond superficial differences and enjoy each other's company—was fairly typical at JPL. There were great stories from days gone by of big parties at the lab and folks moving from group to group to visit, eat, and drink. At that time each section would throw its own soirees, and people would swirl through the buildings getting to know one another better through the buzz of alcohol. But then occasionally people would get a DWI or plow into a tree on the way home, so the lab changed the rules, and parties started happening off-site.

My section had a little holiday celebration at a place called the Athenaeum at Caltech, and a guy named Greg was getting a little loose. He pulled me out into the parking lot and opened the trunk of his car, which was filled with maybe five hundred records. "I know you love rock and roll," he said. "I've given my life to Christ. The music on these records is touched by Satan. Do you want them?" I wondered, did I look touched by Satan? I did not *feel* touched by Satan, but I did accept some of the records. Greg and I stayed friendly, but I always wondered what he really thought of me after that exchange.

Work got more interesting once I got to start working on dynamics questions. Most of the analysis we did in the Structures and Dynamics Group was of the structural variety. When you're analyzing structures, the usual question is "Will this component break under stress?" The answer is yes or no. But when you're analyzing dynamics, the question is more like "How well will this device per-form?" The answers are less binary yes-no and more quantitative and qualitative results.

My first exposure to dynamics came when I was asked to work on one of the last of JPL's big, "grand vision" projects, a $3.3 billion mission called *Cassini-Huygens*. Scheduled for launch in 1997, *Cassini* (the orbiter), carrying the *Huygens* atmospheric entry probe—built by the European Space Agency and meant to land on Saturn's largest moon, Titan—would do gravity-assist flybys, essentially using Venus, Earth, and Jupiter to slingshot itself on its way to explore the Saturn system in 2004.

The first question I was asked to analyze for *Cassini* had to do with how the engines were going to be extracted out of the adapter that connected the spacecraft to the Titan 4 rocket that would lift it into space. The second was the issue of separating the probe from the mother ship.

The separation of the probe required spinning the probe up-ward while keeping it pointed in just the right direction, a bit like spinning a Frisbee very hard while tossing it gently upward on a precisely vertical trajectory. It was a challenging move to execute, let alone analyze.

Around the same time, I started working on another behemoth of a project, *Galileo,* which had been launched in 1989 to study the atmosphere of Jupiter. It wasn't scheduled to arrive at its destination until 1995, but on its way we started to have spacecraft "anomalies." *Anomaly* is space explorer–speak for a failure, a fuckup—hopefully recoverable—in the way the spacecraft is behaving or in your model of how you thought it would behave.

There is a natural stability and elegance to a spacecraft that spins as it travels through space. Like the spiraling of a football thrown by a quarterback, the spinning of the ship keeps it pointing in the right direction, and it does this by natural law rather than through an engineering intervention. That is, if the spacecraft is turned on or turned off, if it is functioning perfectly or having a cranky day, it stays pointing correctly owing to its spin. We always wanted to keep the spacecraft pointing correctly, usually toward Earth, so we could communicate with it, get power to its solar panels if it uses them, and keep the sun heating the parts we expect it to heat. Power, thermal, and communications stability are the golden rules of spaceflight.

But if you want to take a photograph, as we commonly do, spinners kind of suck. It's like trying to take a high-resolution photo while riding a merry-go-round. For taking images and photos, it's better to have a spacecraft that's three-axis controlled. This means that a computer and software and spinning gyroscopes all cooperate in engineering harmony to keep the spacecraft pointed in the right direction, not spinning but with a fixed orientation.

Galileo relied on both. Part of the spacecraft spun, and part of it was pointing in a single direction. With such a configuration, we could take beautiful images and science data from a nonspinning platform and still have the spacecraft be naturally stable.

The rub was that we had to transfer all sorts of power and data from the spinning part to the nonspinning part. To accomplish this connection across the spinning element, we used a large slip ring—metal fingers running along metal bands that allow power and electrical signals to pass from the different halves of the spacecraft. The anomalies were resets of the flight computer that were due, we thought, to shorts or open circuits across the slip ring. To help sort that out, I worked on modeling the fingers sliding on the bands of the slip rings (that is, making a dynamical model of this behavior in a computer).

In the end, we concluded that the short circuits happened on a semi-regular schedule due to the buildup of wear debris from the slip ring finger friction. We calculated that an anomaly wouldn't present itself in the timing of our insertion into orbit around Jupiter, and we could live with that.

I enjoyed all this intellectually, but it still wasn't work that I could bring a lot of myself into. What was missing was what you might call the human factor, where judgment counts and is based not just on what you know but on what you don't know, how well you know what you don't know—on who you are. No matter how much I was drawn to the physical laws because they supplied order, the world of an analyst was too orderly by half. I was, in some sense, bored.

It didn't help that, at this early date, all the work processes at JPL were still huge and complex. The design review procedure made you feel like a tiny gear in a massive machine. It took weeks to get through one of these reviews, with discussion in different departments all around the lab, the review distilled a little more each time, traveling up and up and up into a final meeting in a big room where people like me were never present. There might be a review of the structural design and analysis of a subelement—the main core structure of the spacecraft, for instance—and that review would be subjected to four or five greatly detailed reviews in a room for three days. Meanwhile, similar reviews of the electronic structure would happen separately, after which the findings would be rolled up into a joint review. Then that would become part of a subsequent review of all the mechanical engineering. It was like acting in a movie being shot out of sequence, and you had no idea how each individual scene was meant to contribute to the overall presentation.

It also troubled me that although my answers to specific questions were being used, my more general insights were not being incorporated—they were frankly considered unimportant.

The way we did business, the cognizant engineer (Cog E) designs his structure or device according to his own bent, taking into account the basic functional requirements. Must this thing open once or twice? Does it have to tolerate extreme heat or extreme cold? In this job the Cog E becomes a jack-of-all-trades, master of none, and he may or may not be well versed in structural design. So he may design something that is, at best, difficult to analyze and, at worst, heavy and structurally inefficient.

For *Cassini* I'd analyzed the support structure of a 35-foot-long magnetometer boom that was to hang off the spacecraft, and I'd found that the attachment architecture had fundamentally poor structural attributes. I also thought that the separation attributes for the probe were terrible, and I knew why. If I could see the problems in the finished designs, then the engineers in charge of these elements hadn't been able to see the problems while they were working. It seemed that we were missing a much-needed conversation and collaboration early in the design cycle. There seemed to be a gap between their job of coming up with a design solution and my job of making sure that the solution actually filled the bill. To my eye, there needed to be an advocate for the structural and dynamic essence of the thing while it was still being designed.

According to the established process, the Cog E owned the design, which is the way he liked it, and the analyst didn't have to put his balls on the line, which is the way he liked it. But why not get the analyst in there earlier, to think about the structural function of the device at its inception? Is this thing strong enough? Is it stiff enough? Why go through a three-month cycle asking one kind of question, then another cycle asking another kind of question? Why not integrate the two processes?

I advocated for a more interactive relationship. What if we more strongly connected the roles of cognizant engineer and analyst? This new role would be an analyst in at the beginning. I wanted to call it Cog A, for cognizant analyst. The catch was that, in this new scheme, the Cog E would have to share and the analyst would have to put his balls on the line.

When I proposed to Frank Tillman that we add a new position and new role in the organizational structure called the cognizant analyst, he looked at me as if I were speaking Swahili.

Frank represented the world of Southern California aerospace that had come of age in the fifties. He had worked for Boeing and for many others and had contributed enormously. But my sense was that he had never in his life tried to change the way anything was done. He just tried to do things well, the way he was asked to. He didn't even know whom to talk to about a better way of organizing the work.

As I look back on that frustration, I can see a rich cocktail of youthful naiveté, headstrong certainty, and a dash of genuinely useful observation about the way the organization operated. I've since learned that this mix is fairly common in intelligent young employees first stepping into a job for a big firm, be it JPL or another aerospace firm, Apple or Google, or anywhere else.

A fresh perspective frequently brings with it insight not available to those on the inside. Zen Buddhists call it *shoshin*—"beginner's mind." That is one of the greatest gifts an entering employee can give, but it also presents a challenge. The same vision that sees the holes in the operation does not see why those holes might have been accepted long ago as a necessary evil or even put there by an act of will. Beginner's mind presents both challenges and opportunities for the institution and for the new employee.

The institution must listen, and listen with an open mind. Frequently these youthful observations, like mine about inviting the structural analyst into the design process, are awkward, sometimes not well framed, and easily dismissed. The challenge for management is to look past the warts and imperfections of the delivery for the essential virtue that might lie in the suggestion. In my example there was a definite need for those insightful early conversations about structure to be folded into the design process. Indeed, we have those conversations in JPL teams today. But we have them without needing to create a new role or job title, as I had suggested to Frank.

In situations like these, the challenge for the young employee is patience. The media is rife with examples of overnight tech moguls, which makes it all too easy for a new employee to think: "I have been here two years, and I don't see my mark on the institution, my advancement in authority, or my jump in compensation. It's time for me to be recognized or leave."

Here's what I wish I could tell my younger, impatient self: Advancement and influence in any industry do not in general keep pace with an industry's most famous outliers. At least in a meritocracy, such as JPL, if you do good work and really focus on mastery and excellence, good things happen, for the institution and for you. That is not to say that "the institution" is a benevolent, all-knowing, and all-rewarding entity, but any institution desperately needs good people. There is a vacuum at the top. Their desire for good people and talent is insatiable. If you do good work that is valuable to the institution, you will inevitably be vacuumed upward.

What's more, true authority comes not from a title or position but because your words are well thought out, or at least strive to be. Nothing puts more weight into your opinion than that it is well considered, well articulated, and coming from a high performer. If you want people to take you seriously, better put your head down and be useful to them.

Luckily for me, Frank was tolerant of my youthful agitation, and I wasn't canned for being a pain in the ass. But what really saved me was the opportunity to engage in a more challenging problem: *Pathfinder*.

At this same time in my career, while I was working on colossal projects like *Cassini* and *Galileo*, I was asked to do some dynamic simulations of the parachute, back shell, and lander for a smaller mission called *Mars Pathfinder*. Scheduled for launch in late 1996, *Pathfinder* would carry a small rover called *Sojourner* to the Martian surface.

Pathfinder had a kind of Skunk Works, underground aura to it. Although my involvement was very limited, I attended a few of the team's meetings a half-dozen times or so, and I could sense that they were doing things differently. Standard operating procedure in most of the engineering groups was to keep your head down, do your analysis, and remain stoic at all times. But the *Pathfinder* team let their feelings fly. Meetings could erupt into shouting matches, but then everyone would walk out of the room unfazed and on good terms. They also seemed to have pared down the decision-making process, especially the review process. They didn't document anything beyond essential communication—no paper trail for the record or for NASA higher-ups.

John Casani, formerly chief engineer at JPL and one of the giants of space exploration for more than fifty years, explained to me once that back in the fifties at JPL, they used to say, "Here's your drawing, but it won't do you any good without the engineer stapled to it." The guys on *Pathfinder* seemed to have recovered that sense of personal connection, investment, and contribution, as if each of them was stapled to whatever design he was advocating. With *Pathfinder* it wasn't a matter of required drawings going through a series of administrative hoops; rather, it was a return to the passion and commitment of an engineer standing right there, figuratively if not literally, as the machinist made the component.

Once I was walking by a conference room when I caught sight of Dara Sabahi, the mechanical systems lead for EDL, engaged in a debate with other members of the *Pathfinder* team. I remember the language and tone more than the content. At one point Dara said, "That is complete bullshit, and I won't let that happen." To many readers that may not seem like very strong language, but for me, coming from an environment where I barely met the people I was working with and everything was filtered through others, this sounded like war. It was the kind of war I wanted to be part of.

I heard mutterings and complaints from the old guard that the *Pathfinder* team was cutting corners. But to me it sounded like the spirit of the old JPL I'd heard about.

Having gotten a taste of this more freewheeling, put-your-heart-into-it approach via *Pathfinder,* I was ready to get out of the slow-moving waters of the stodgy, huge missions and into some of the streams where those more creative currents were starting to flow again.

Ruthann was finishing up her master's at the time, and we were making plans to get married. She was looking to shift her math and physics training toward a doctorate in finance, and after our wedding she began applying to grad schools all over the country. I wanted to stay at the lab, but by this time I was ready to get back on the path toward a doctorate myself, and I began looking into graduate programs in engineering that would offer us a spousal arrangement, two for one on the same campus.

The compromise we found was the University of Wisconsin–Madison. They admitted Ruthann to their finance program and gave her money, and their engineering mechanics department admitted me. I remained an employee of JPL, working there summers and holidays, and got a job at Wisconsin as a teaching assistant, which there meant that you weren't just some sort of proctor but you actually taught the courses.

We moved to Wisconsin in the fall of 1994, and for the next five years, we jockeyed back and forth between Madison and Pasadena. I was surprised to discover that I loved teaching. It was terrific fun to engage with thirty-odd students all thinking about the same concept—some nailing it and some getting it wrong—and trying to identify and communicate the essential lesson in the material. It was through teaching that I first learned how to hold on to the doubt with others, not just alone.

Take a typical professor-student interaction. One day during my office hours, a sharp student—let's call him Tim—walks in with an assignment that is neat, tidy, well written, looks right, but is somehow wrong. The assignment: the classic bowling ball problem.

A popular exercise in undergraduate physics classes, this involves asking questions about the dynamic behavior of a bowling ball as it travels down the lane toward the pins. Everyone who has ever bowled knows that the ball starts out sliding but eventually starts to roll. This behavior seems frighteningly complex, and that makes it a great and challenging problem for college students.

Tim comes in with this problem, and he's solved it. At least he has the correct answer, but he hasn't used the proper approach, the classic method, which I recommended. So now he's come to my office to tell me that I'm wrong.

To solve the bowling ball problem, you need to utilize Newton's second law of motion in two dimensions: the kinematics of the eventually rolling ball and the conditions of the ball when initially released. For undergraduates, that is a lot to untangle. They often neglect at least one of the puzzle pieces.

This is the case with Tim. He's very bright, headstrong, and certain of his correctness. After all, his solution matches the answer in the back of the book.

The easy response is to tell him he's wrong, hand him the correct working of the problem, and push him out the door. That technique minimizes my time in the presence of both my doubt (am I really correct?) and his (does this guy Adam really know what he's talking about?).

The other approach is to wade into Tim's work, line by line, term by term, not knowing if I will discover what went amiss; not knowing if I can prove to Tim that I know my stuff; worrying, maybe, at some deep level, that I don't.

True learning lies in the latter approach: to embrace the confusing mess of tangled thought and tangled algebra and let curiosity loose to hunt for what is right and true. This is not a process without anxiety. It requires that both parties sit side by side with doubt and hold on to it no matter what the ego demands.

In the end, it was a delightful case of two wrongs within his work making a right, or an apparent right. After wading through the problem together, Tim and I agreed on the primacy of Newton's second law and the importance of going slow in our work.

Many years later, Tim has a very successful career.

Holding on to the doubt (HOTTD) is a big deal, and it is not always easy, but it has to happen, and it has to happen at many levels. Not only does HOTTD make you a better thinker yourself; it is indispensable in making a good thinking team. Developing something new and novel is a nonlinear process. Being part of that process, as a team member or a leader, can be anxiety provoking, because we don't know where it is going and we don't know if we are really going to find a solution that works or meets our needs. The temptation to short-circuit the team process is strong. It is only by HOTTD that we can allow ourselves to fully understand the problem we are solving and fully develop the solution.

In my current job, we're way beyond the bowling ball problem. I don't have answers beforehand, and I don't know the solution method that will work. But I still use the same process I developed back in Wisconsin whenever I'm leading a team through an engineering challenge. I've learned to be comfortable sitting beside doubt, and that I've got to help the team feel that way, too. Together, thus far in my life, we always find a solution.

Chapter 4

SELF-AUTHORIZATION

ONE OF MY FAVORITE TEDDY ROOSEVELT QUOTES IS "FAR and away the best prize that life offers is the chance to work hard at work worth doing." I feel very lucky to think that the work I do is worth doing. I also feel very lucky to work on a scale that requires people to work together in teams. I appreciate teamwork so much that I've developed a riff on Mr. Roosevelt's quote: "That great work requires many people coming together is one of the great prizes life has to offer."

To really participate as a member of a team, you have to bring yourself to the process. Bringing all that you have requires offering up your opinion in the absence of an invitation. It requires self-authorization. You need to believe that you have the answer, and you need to give it to the team, even if you only *think* you have the answer. It is a form of leadership, and it is needed at every level and in every element within a healthy and high-functioning team.

At JPL we use a matrix organization structure—a blend of line and project organization—for the management of the work. Everyone is assigned one or more projects, which last for three to seven years and culminate in a mission, like a spacecraft going somewhere and exploring. Simultaneously, everyone is assigned to a group or section that specializes in a discipline or type of task, together making up the line organizations. For example, we have a Mechanical Engineering Division with about a thousand people in it; it has three sections, and the sections have groups within them; each group may have five to twenty people.

Over the years, how we assign work has evolved. When I first arrived, you got your job from your group supervisor, who received the work in a coordinated way from the section manager, with input from the division manager. An old-school project like *Cassini* would have the mechanical work assignments managed by the line organization. The project would say, "Hey, Mechanical Engineering Division, do us some mechanical engineering," and then the line organization figured out who would do the work and when and how it would be done.

Today we use what we call soft-projectization. We still have the line organization, but the work is assigned more directly (though not completely) by the project. In this arrangement, a project leader comes to the line organization and says, "I want Jane Smyth to work this specific mechanical engineering task for the project." This creates dynamic tension between the line and the projects, with projects competing for key staff and the line trying to balance the lab's needs across a range of projects.

When I finished my first year at Wisconsin and came back to the lab for the summer of 1995, JPL was in the shift between these two modes. I was looking at another stint in the fairly dull (for me) domain of Frank Tillman's Structures and Dynamics Group, being assigned work from Frank as needed in an old-school kind of way.

At the first meeting of the group that summer, I got to talking to a woman named Ann Mauritz, who was administratively still in Frank's group but had actually made the jump into the newer mode, moving from being an analyst to being a Cog E and engineering manager. In the project world, she was serving as chief engineer in the early development stage of a new mission called *Champollion*.

Named after Jean-François Champollion, the French scholar who helped produced the first accurate translation of the Rosetta Stone hieroglyphs in 1822, *Champollion* was an attempt to land on a comet or, more precisely, to bring a spaceship in proximity to a comet and link up with it by driving a spike into it. To me, harpooning a comet sounded like an interesting problem in mechanics and plasticity. I told Ann I'd been doing some plasticity theory work in Wisconsin, and she said, "You should go talk to them."

We both looked over at Frank, who gave the nod, and so I went.

JPL is filled with directorates—like Engineering and Science, or Astronomy—which are broken down into divisions, such as Mechanical Engineering or Telecommunications, which are further broken down into sections, like Section 354, Applied Mechanics Technology, which was where I lived. The comet stabbers were 352, Mechanical Engineering. Both these sections lived in Division 35, Mechanical Engineering, but the guys in 352 were the tinkerers, the designers, the Cog E's, and we in 354 were their egghead cousins. Usually we would have been handed a problem late, after the design had been done. So it was considered pretty odd for a 354 guy to walk into a design meeting in 352.

"Ann said I should drop by . . . ," was the best I could do to explain myself.

The modern matrix structure at JPL allows job assignments to be very fluid, kind of like the Wild West, especially in the early stages of a project. People are encouraged to try to stretch into roles that may be within their grasp. Even if they're not up to those roles, at least they get the chance to try out for the part. It's very Darwinian in the sense of fostering experimentation and variation, but also Darwinian in the sense that a development-stage project head, sometimes called the proposal manager or the preproject manager, is often replaced as the demands of the job become clearer. So this trying-out business can be humbling, and it requires a strong ego that can take the ups and downs.

Some jobs require that a requisition be written and approved before someone can be hired. That would have been the case for Ann's role as chief engineer. For lesser jobs, when someone simply needs a worker bee, which would be the case for my presumed role on *Champollion,* the only paperwork needed is a work authorization memo (WAM), so that the hours can be charged to a budget line. That's what Ann did for me—she WAMmed me.

The team already trying to figure out how to harpoon this comet consisted of Randy Lindemann and Chris Porter. Both were very good Cog E's, around my age, which is to say early thirties, and at my level, which is to say master's, not Ph.D. Randy had studied at the University of Texas in Austin. Chris had grown up in rural Louisiana, where he and his father had hunted rabbits and possums for food. His background made him nothing if not practical, which offered a useful contrast to Randy's slightly more book-smart perspective. They were both all about hard data and designing mechanical devices, and both possessed a limited tolerance for anything flamboyantly theoretical.

Now, "landing" or anchoring onto a comet is difficult. A comet is just a hunk of ice and dust hurtling through space, and there is really no gravity to speak of. Things don't stay put on the surface, and the slightest bump or jostle and you're floating away. Take the *Philae* lander, in 2014 the first payload to ever attempt to land on a comet. It bounced—twice, the first time shooting more than half a mile back into space—before settling feet-side up (not good), on the wrong side of the comet, when the screws designed to hold on to the ice failed to work. But fifteen years before *Philae,* when I was working on the problem, we didn't even know what the surface of a comet looked like, rough or smooth. The accumulated wisdom of science held that the material it was composed of had a strength that ranged from hard concrete to the softest, fluffiest snow.

The *Champollion* engineering team had chosen to drive a single telescoping spike into the surface. If the surface was hard, only the tip of the spike would penetrate. If it was soft, the whole length would penetrate. It was a sort of self-adjusting method.

Randy and Chris were trying to get a handle on basic penetration mechanics by dropping pointy things that looked like small steel missiles off the tallest buildings at the lab. They were developing their basic intuition, a necessary if time-consuming process. They'd design something, then test it, then design and test something else, relying on a sort of a guess-and-check algorithm.

My head spun when I looked at the breadth of the problem. It was hugely complex. As Chris and Randy worked to understand the penetration problem, it was not at all clear how we would ever converge on a solution. We hadn't broken the problem up properly and divided it up among ourselves. What things did we have to test? What could we understand from basic physics and analysis? What pieces did we need to engineer? And what were the central questions we had to answer? There seemed to be a vacuum of technical leadership in the way we were attacking this.

I then did something I had never done before, an act that perhaps more than any other changed my career and perhaps my life. I stood back and allowed myself to look at the whole problem of anchoring to a comet. I then broke the problem up into the pieces that seemed logical and useful, pretty much just as I had been doing with my students back in Madison. And then I, the dweeb from 354, opened my mouth and told Chris and Randy, the comet stabbers from 352, what I thought.

I started from the perspective that I was definitely *not* the boss or anything like it but, rather, that I had thought about our problem and broken it up into the pieces that we could analyze, the pieces that we needed to test, and the data we needed to get out of our testing. I shared my reasoning with them, and they agreed to follow the plan of work I proposed. The lab has a long-standing tradition of yielding to the strengths of arguments, but this was the first time I had stepped up and offered my arguments for consideration.

It so happened that Dara Sabahi, the man with a passion whom I had seen arguing during the *Pathfinder* mission, was the guy in charge of this whole comet lander study. When he returned from vacation and saw what I had done, he put me officially in charge.

Dara was an Azerbaijani Turk from Tehran who'd come to the United States as a teenager. He'd begun his career as a stress analyst with a contracting firm, but it turned out that he had great gifts as an intuitive leader, like an old-world tribal elder. Dara gets his teams to actually do what they say they're going to do by finding out the team's deepest concern and then echoing that back to them in his directives. Naturally they do what he's asking of them, because it was embedded in their desires in the first place.

I wasn't making any particular claim to leadership, nor had I been given prior authorization to lead. I'd acted like the team's leader by asking Chris and Randy questions that would help them organize their thinking and solve their problem. That's what Dara liked. His promoting me endorsed the idea that you have to rise to the performance level of the job you want to have, which means that you have to just freakin' do the job. If you come in acting like the boss, bossing people around in order to feel powerful, the most common response is going to be "Who the fuck are you?" So the trick is to exhibit leadership—lead—without having to claim leadership or to subjugate others. This is leadership really as a service function, as a gift to the group.

I used to think of leadership as the sign of power. At JPL I learned it is also a sign of service. Think pit crew chief or team captain. These roles serve only to facilitate the performance of a collective group. Someone has to take the role of intellectual coordinator and Mother-Goose the problem. If you offer your direction to the group for their endorsement, you are not imposing your will. You are simply gifting your thinking of how the group should attack the problem. The converse lesson is that anyone who waits around to be authorized to speak up is never going to be a leader. It's just not in their nature.

This promotion was hardly a great leap up the JPL chain of command. I would be coordinating the part-time efforts of a team of two—three, counting me. Randy and Chris were fine with me leading, because I did very little of it. They had other jobs, and my leadership held sway over a mere fraction of their time. And my "management" responsibilities would be light enough that I could handle them after I returned to grad school, staying in touch by phone and flying out now and again.

The more time I spent on *Champollion,* the more I was exposed to intellectually challenging conversations. Dara and Brian Muirhead, the project manager, felt free to engage me in technically specific debates all the time. One day in Brian's office, we were going over a particular point when he broke in abruptly with "Well, what about the sublimation!?" (This refers to a kind of melting that in the vacuum of space moves directly from a solid to a gas. He was making the point that our spike would necessarily be much warmer than the comet.) His intonation was almost an *en garde!* It certainly was not polite and deferential, nibbling around the issue. It was more like punch and counterpunch. I liked it!

"Oh, we're going to use fingers," I said, referring to the curved-leaf spring-like steel tines that would stretch out and keep contact with the surface as the comet material receded from the spike.

To which he replied, "Well, are you just going to rely on the heat path along the spring length?"

And so it went, punch and counterpunch, back and forth.

These direct, assertive, full-on encounters made the other groups seem passive and risk-averse by comparison, as if they feared the truth and the changes it could bring. These *Champollion* guys, who had been the *Pathfinder* guys, were more like the naked barbarians I'd heard about as part of the foundation myth of JPL, brawling for the truth around the campfire. These were the engineers stapled to the drawing.

Another of these full-on, committed guys I got to know on *Champollion* was Miguel San Martin, an electrical engineer with an expertise in guidance and navigation. Right away I could see that he was incredibly insightful and smart. It was only later, on subsequent missions, that I would realize how powerful his mind was.

Miguel is tall and aristocratic looking, so much so that in a suit he could pass for a government minister in some sophisticated European capital. He's also an intellectual, whose conversation often begins with "I was just reading this piece in the *Atlantic*. . . ." He's funny, too, often puncturing the tension in a room by coming out with an imitation of Ronald Reagan's breathy "Well . . ."

But Miguel's greatest virtue is a capacity to bring a practical sensibility to a highly mathematical field and make things happen with the least amount of fuss, and research, and lines of code. He strips it all down to its essential, practical core. We didn't know then that Mig and I would become friends and partners through some of the greatest moments in our careers.

I went back to Wisconsin for the school year, and by the following summer, *Champollion* had reached the point when NASA would have to fish or cut bait—make a solid commitment and really start investing money, or kill the project. Because of changes in funding priorities at headquarters, they chose to do the latter. Which did not stop JPL from coming back with an even more ambitious project, which nonetheless could restart the clock on the development process and go back to burning money at a slower rate.

This replacement project was called the Comet Nucleus Sample Return (CNSR) mission, and the idea was to combine *Champollion*'s basic concept—anchoring to the comet nucleus to do surface science—with the additional wrinkle of bringing back material from the comet for analysis on Earth. The heightened aspirations increased the budget but offered a far bigger value in terms of scientific bang for the buck. I was asked to take up the same position I'd had on *Champollion,* which was landing and anchoring lead.

Dara was again the man in charge. This being summer, he was at his cabin in the Sierras, and Miguel and I worked on the CNSR effort together.

The mission design called for a solar electric propulsion approach to getting to the comet. That technique would require very large solar arrays to provide the needed thrust out where we planned to rendezvous with our target. Landing with such large arrays presented a daunting challenge.

Before he left, Dara had made it clear that his idea was to depend on a single vehicle throughout the mission: on the way to the comet, landing, and then flying home. But while he was away, we decided that the best thing would be to have a second vehicle—a daughter ship that would attach to the comet, then redock with the mother ship for the return. The daughter ship would use batteries and land without the clumsy and flimsy solar panels.

At our first meeting after Dara got back, he was like a dad who'd left very specific instructions for how you were supposed to stack the cordwood . . . and then you'd decided to improvise.

We were trying to explain our thinking, and Dara was getting more and more angry. At some point I was trying to defuse the situation and said, "Look, Dara—"

At which he said, "No. You look. I don't need this," and stormed out of the room.

Everybody who saw it was stunned. He was like the guy who pulls the pin on the grenade with the handle down and says, "Do what I say or else!" His storming out was like letting the grenade explode. It's upsetting to confront that kind of emotional brinksmanship; even more so to go over the brink! To me it felt like I'd seen what too much emotion could look like.

We solve problems at both the conscious and subconscious levels. Dara was and is a naturally intuitive leader and problem solver. His emotion was there to protect the space he needed in order to problem-solve in a kind of subconscious, perhaps even slightly irrational, way. In an engineering environment, this combination of the rational and the irrational is so unusual that it shakes people up. That makes it an incredibly powerful catalyst for getting at the truth, but like any powerful ingredient, it needs to be used sparingly.

By the time I was working on CNSR, I was also wrapping up my Ph.D. thesis on the inverse structural response problem. (If you were to take a building, let's say, and shake it, which puts forces on it, and then you measured the response, that's the *forward* structural response problem. The *inverse* problem is to take the responses that have occurred and try to back-calculate the forces that were applied to achieve that response.)

In writing a doctoral thesis, I was reaching the end of the line of our formal educational system, but the lessons I was absorbing through exposure to people like Dara and Brian and Mig were going to be just as valuable going forward.

In the land of the rational minimalists, I found myself thinking more and more about an essay I'd read in an English class back at College of Marin. Written by a Harvard teaching assistant, it told the story of an undergraduate with time on his hands who followed a friend into an exam and took the test just for fun. On the essay portion, the test crasher got an A; his friend, who was actually enrolled in the course and had done all the reading, got a C.

The difference was that the friend, who had all the facts, didn't do much with them. Whereas the test crasher, who had no facts other than the few suggested in the way the question was posed, spun out an amazing exploration of what he assumed the issues to be.

The teaching assistant telling the story came up with two verbs to identify the different approaches. The first was "to cow," which boiled down to presenting facts as a substitute for understanding. The second was "to bull," which is to present evidence of an understanding of form, context, and frames of reference in order to suggest a nonexistent familiarity with the facts. Cow is data without a model, theory, or connections. Bull is a model with no data to support it.

You can't bull your way through life, and you certainly can't bull your way through engineering. But you can't cow through it, either. Essentially the "cows" are the facts that link us to the observable universe and the "bulls" are the connections between the facts, the causal relationships between the observations. The bulls are our models of the universe—they are what we use to predict as-yet-unobserved cows.

I have come to realize that, in the business of building robots to fly into space and explore on behalf of humanity, most of our problems arise from a lack of bull. Rarely do we lose spacecraft because we lack facts or because a particular fact was wrong. We lose spacecraft because we did not appreciate the links or connections between the facts.

When *Mars Polar Lander* was lost in late 1999, one of the leading candidates for the cause was that a contact switch in the feet of the legged lander was triggered by a jolt associated with the deployment of the leg. This occurred some 80 meters (262 feet) above the surface of Mars. Bummer. So close! The basic facts were understood by the team, but the connection that was not appreciated was that the test to ensure that this premature triggering would not occur had been conducted improperly, and that when folks considered re-testing, they decided not to. There was a potential flaw in the contact switch and a possibility that software reading that switch might fire the engines when the craft was 80 meters above the surface. The smoking hole on the surface of Mars where a lander should have been is a testament to the trickiness of finding all the necessary connections—all that bull.

Constantly striving to find and understand those connections is the prime motive for science, but it is also critical in engineering or in any other practical enterprise, including any business. It is not enough to log and tabulate the observables; it is essential to develop an actionable understanding of the underlying causes of the observables.

In August 1999 I was awarded my Ph.D. Ruthann finished her doctorate at roughly the same time, and she got two job offers: one with J.P. Morgan in Manhattan and one as a professor at California State University, Fullerton. She had no real interest in being a teacher, and she liked the idea of Wall Street, so I said go for it.

We assumed that she would spend maybe a year in New York and then come back, and we decided we could manage the bicoastal thing for that long. The truth is, there were issues with our marriage that had been present from the beginning, and we were struggling. I think the separation was a way of not having to address those issues directly.

She got an apartment in Brooklyn, and I helped her move. I also arranged for my job with CNSR to be one week on-site at JPL and one week working from New York. But then CNSR got canceled, so the bicoastal thing was reduced to my working in California and spending every other weekend in New York.

I was no longer seen as an analyst, no longer a part of Frank Tillman's group, so I began to poke around for new opportunities at the lab. One new project I heard about was called Mars Exploration Rover (MER), which was going to send up two small, robotic geology labs called *Spirit* and *Curiosity*. The idea was to use these rovers to explore various rocks and soils to try to determine the nature and extent of past water activity on Mars. To get the maximum bang for the buck, the rovers were going to be dropped at two very different sites: the Gusev Crater, thought to be a former lake bed in a giant impact crater, and Meridiani Planum, a broad plain where the mineral deposits suggested a history of water.

To overcome our understandable insecurities about the challenge of getting safely to the surface (we had just lost *Mars Polar Lander*), we would use the same landing system we used for *Pathfinder*. We would just cram more spacecraft into the same system.

Even with these ambitious scientific objectives, management at JPL saw MER from a design and operational standpoint as simply a reflight of *Mars Pathfinder,* the archetypal "faster, better, cheaper" success that had landed *Sojourner.* So they staffed MER quickly with whoever was readily available. By the time I got back from New York and offered my services, I was too late to get in on it.

A few months later, though, management realized that this was going to be a much harder job than they'd thought at first, and they brought in Dara to help. Put in charge of all the mechanical engineering for the project, he asked me to come sit alongside Greg

Davis, the mechanical engineering head for Entry, Descent, and Landing. Dara was never explicit about this, but I believe he saw that changes to the design might be needed, and he had a hunch that, given past experience, I might appropriately self-authorize to look for those changes.

An organization needs to reflect the people in it. We are not interchangeable human units of effort or skill. Each of us is unique, and the organization needs to be shaped around us.

Not wanting to be held hostage to talent that might walk out the door, most organizations chafe at the idea that people are unique and that any one person's contribution might be crucial. Institutions want to believe that policy and process, not individual belly buttons, determine the quality of their product. Thankfully, JPL, more than most, accepts that policy and process without the right people is a losing proposition. I think Dara knew this, and I think he decided to deploy a different human tool (me) when the new and different challenges facing MER became evident.

Of course, it was unsustainable for both Greg and me to try to do the same job. Eventually the organization morphed to put me in the role of mechanical systems engineer for EDL, and Greg in charge of all the verification and validation not just in EDL but across the entire mechanical effort, the rover included. In these re-defined roles, we were each positioned to contribute in a maximal way. The organization structure was reengineered to reflect the skills and capabilities of the people within it.

I came back to JPL not at all clear about what I was getting into. Ruthann continued to live in Brooklyn, and I went out to her once a month and she came out to California once a month. The thread binding us was getting thinner and thinner, but even so—or maybe as one of those last-ditch efforts, like a dying fruit tree sending out more blossoms—we started trying to have a kid.

After six months' effort, we got pregnant. Nine months later, on October 10, 2002, our daughter was born, and Caledonia was her name.

They say it takes a while for a baby's eyes to focus, but when Caledonia first pinned me with her gaze, it was a revelation. There was no duplicity there, no hiding behind a veil of politeness or fear. I realized I no longer wanted to hide myself from the judgment of others. I had had an opportunity to step up into leadership in my work, fully there and fully committed, and now I wanted to apply those lessons to my life. It was as if taking more personal risk at work had been practice for taking more risk in my personal life.

Also, the more comfortable I became with putting myself on the line in my personal life, which meant eventually facing up to the dysfunction of my marriage and choosing to leave it, the braver I became at work, sharing myself more fully and connecting more deeply with my colleagues. Landing a spacecraft on Mars was going to require a lot of technology that was tantamount to magic. But it was also going to require a lot of human beings with the kind of trust in and commitment to one another that might be even harder to come by.

In hindsight, that first act of self-authorization on *Champollion* set me on a path that transformed how I understood my personal life, which in turn changed how I approached work. I guess I am a late bloomer and was in need of training wheels. I had used school as my first set, at College of Marin, where grades and class performance was like "practice life." Again, now stronger, on *Champollion*, that first leap of committing myself, leaping and offering what was on my mind, instead of just calculating the numbers for someone else. With each of these steps, I committed more, to life, to the present—risking more. I would bring myself more fully to my life, both professional and personal. My work was changing me, and I in turn would try to change it.

Chapter 5

SYSTEMS ENGINEERS

WHY GO TO MARS?

On one level, the answer might be the one that George Mallory, the mountaineer, gave when someone asked him why he climbed Mount Everest: "Because it's there."

Mars has undoubtedly been there, visible to the naked eye from Earth, for a lot longer than there have been members of our species to look up and see it. The only objects that loom larger in the sky are the sun, the moon, Venus, and Jupiter. And just the look of it would have been enough to make it an object of fascination to our ancestors. Iron oxide—rust—in the soil makes it appear red, which caused many cultures, from the Chinese and Hindu to the Babylonian and Greek, to associate the planet with fire and the powers of war. Thus its name, Mars, after the Roman god of war.

But more important, Mars is very much a sister (brother?) planet to Earth. It's the next one over, the fourth from the sun, with half our diameter and half again as distant from the sun as we are, meaning that it gets only 43 percent of the daylight we get. Its rotational period and seasonal cycles are similar to Earth's, as is the tilt of its axis. It also appears to have been formed at more or less the same time as our home planet, roughly 4.5 billion years ago.

These similarities, along with the presence of carbon, sunlight, water, and nitrogen, contribute to the most intriguing single thing about Mars, which is that it's the physically closest place we can look for extraterrestrial life. The next-best options are probably Saturn's moon Enceladus and Jupiter's moon Europa, and then it's off to the Tau Ceti system, seventy trillion miles away.

Admittedly, talk of life on Mars has been consistently overheated. In the 1880s the Italian astronomer Giovanni Schiaparelli observed what he thought were linear structures on the Martian surface and referred to them as *canali* (meaning "channels"), which prompted Percival Lowell, the nineteenth-century intellectual who founded the Lowell Observatory in Flagstaff, Arizona, to misinterpret the name and speculate about the possibility of a lost canal-building civilization. In 1901 the Serbian-American inventor Nikola Tesla wrote about detecting, in his laboratory, sounds from another planet and the possibility of communicating with Mars. With those bits of speculation, the popular imagination was off to the races. H. G. Wells wrote *The War of the Worlds,* which spawned Mars books by Robert A. Heinlein, Ray Bradbury, Lester Del Rey, and Philip K. Dick. Life on Mars became a staple of science fiction, and human-Martian interaction served as the central plot device in films from Thomas Edison's silent classic *A Trip to Mars* to *Total Recall.*

Obviously the kind of life we might hope to find on Mars has nothing to do with little green men or dudes who look like Arnold Schwarzenegger. Microbes, even fossilized traces of microbes, would be an astonishing discovery. To me, life somewhere other than Earth would be a profound and comforting discovery. Profound because it would mean that we are not the only living place but part of a larger community of living places. Comforting because it means that if we screw up and ruin all life on Earth, we would not have destroyed the only living things in all existence. Scientifically, the discovery of life on Mars would vastly help us understand the origin of the solar system and the origins of life within it, and whether the first spark occurred on Earth or someplace else, like Mars, or in several places all at once. These different scenarios themselves have implications for how likely it is that life flourishes in other places.

It's sexy stuff, and as the Mars Exploration Rover mission got under way in 2000, JPL had owned the planetary-exploration game for decades. But just then the whole Mars program—what you might call the jewel in JPL's crown—was threatened with extinction.

That's because of the double disasters that occurred in 1999 and the desire, on the part of NASA headquarters, for increasing competition within its own ranks and within the ranks of its civil-space industrial contractors. It started like this: After the *Mars Pathfinder* mission (launched in 1996), which cost about $350 million, our frequent contracting partner Lockheed Martin offered to build two spacecraft, an orbiter and a lander, for about that same price. So the next test of "faster, better, cheaper" became an attempt to build two spacecraft on the same schedule and at the same cost as *Pathfinder*. It was a limbo dance, and we were setting the bar lower. Could Lockheed, with our help, go that low?

These spacecraft shared some similar parts but had different missions. They were to be launched within a few weeks of each other, putting one on the Martian surface (*Mars Polar Lander*, discussed earlier) and one in Martian orbit (*Mars Climate Orbiter*).

Further economies would be gained by reusing off-the-shelf designs, some of which, like the designs for the aeroshell (a rigid shell that protects the spacecraft from heat and pressure during entry into the Mars atmosphere, descent, and landing on the planet) and *Polar Lander*'s legged platform, dated back to the *Viking* missions of the seventies.

But while embracing risk to come out ahead on cost sounds like a reasonable bet, one that had worked for *Pathfinder,* the economy doesn't hold when you take two shots on goal and miss them both.

Orbiter failed because of simple confusion over units of measurement—ouch, that hurt. Our agreement with Lockheed Martin stipulated that the units were to be metric, but in one column of the small-forces file—small forces being things like solar pressure—the values were registered in imperial units (pounds). Trouble was, we simply didn't have enough people working on the job to scrub all the data completely clean. So after the file was transferred to JPL and it was time to send the craft into orbit, the discrepancy put us too close, and the spacecraft burned up in the atmosphere. A simple mistake—a bookkeeping error, really—with huge consequences. Naturally we were the fodder of the late-night talk show hosts: The brainiacs can't get the units right! It felt like everyone was piling on.

JPL scrambled to find out what went wrong, trying to save *Polar Lander,* but we lost that, too. As mentioned earlier, the most likely scenario was that as the legs deployed during the last stages of descent, the computer mistook a mechanical jolt for touchdown with the Martian surface and shut the engine off, which allowed the lander to plummet those last 80 meters (262 feet) or so. That would have been caught in the botched test I mentioned, but we didn't realize how important that test was to redo. Once more we simply had too few people to do all the checks and double checks needed to connect the dots and get everything right. So two separate mistakes, two separate disasters. The limbo bar was too low, and we were groaning, lying flat on our backs in pain.

"Faster, better, cheaper" had embraced the trade-off between slight increases in the risk of failure and the promise of making a lot more exploration happen. Some unknown point had always existed at which the terms of the trade-off were no longer acceptable, but you never knew until you got there. *Mars Polar Lander* and *Orbiter* had found that drop-off point, and now everyone was painfully aware that with the Mars Exploration Rover mission, both NASA's credibility and JPL's were on the line.

Even so, the initial plan for MER still embraced the idea of getting by with less and reusing past designs. The "faster, better, cheaper" playbook for MER was to take two much larger rovers, fit them into the preexisting *Mars Pathfinder* Entry, Descent, and Landing system, and send them up on two separate flights just a few weeks apart.

The "this is just a reflight" mind-set had led to a "build to print" game plan, meaning that we'd be building off the old designs or blueprints. As a result, the lab hadn't staffed the project with architects—that is, the kinds of engineers who can look at a blank screen and begin to envision something radically new, or even question the plans in front of them. Line management had simply brought on engineers who were the very best at bolting shit together and testing it again and again until they were sure that every component was perfect.

But reality soon caught up with us. At every turn we discovered that we were going to have to make changes we hadn't counted on. Ultimately we realized that in order to succeed, we were going to have to blow up the entire notion of build-to-print and actually create all new components. Given that EDL components are all interconnected, to one another as well as to hundreds if not thousands of other components throughout the spacecraft, EDL for MER became a story of wave after wave of changes, each new wave adding to the pressure.

And we didn't have much time. Depending on the conjunction of their orbits, Earth and Mars can be as distant as 249 million miles or—once every twenty-six months—as near as 33.9 million miles. To get there with the rockets available, we had to wait for just the right time. We had just missed the window of opportunity for this cycle, so the next moment of proximity would be 2003, and twenty-six months was very tight, even if we'd been sticking with build-to-print.

One benefit of being on a such a schedule is that it's hard to spend as much money. Human beings and communication among human beings are where the real money goes, teaching and learning as you scale up from one person to three people to a thousand people. With limited time there's a maximum slope of expenditure bounded by the limits that time places on human interaction. And yet, as I was coming to realize, in building spacecraft as in most other complex endeavors, success or failure depends on those human factors—the quality of interaction, the clarity of communication— every bit as much as it does on the technology.

At the lab we spoke of being engaged in a battle for our lives.

A single, decent-size rover, kind of shoved into the landing system of the *Mars Pathfinder* mission, was going to be a tight fit, but NASA and the lab felt we needed a sure bet. *Pathfinder* had been a success, so why not reuse that system?

Soon we were asked to build a second copy of the *Pathfinder* system, which increased the chance of success and offered certain economies. If you fly two rovers, chances improve that at least one is going to make it. And the cost of the second copy is pennies on the dollar if you build it at the same time as the first.

That said, we had a *lot of work* on our plates. Achieving the kind of quality and clarity we needed under time pressure was going to be rough. So while I think the coldly rational, engineering part of each of us was saying, "Of course it will work," the human part was saying, "Maybe not."

I liked being in the middle of an intense, high-stakes storm, but I also felt how easily it could sweep me away. Some people handled the pressure with grace, and others folded. I aspired to be in the former camp, so I fashioned an approach: Whenever I felt I was being swept up in the storm, I visualized being in a warm coat on a snowy slope with a blizzard swirling around me. I would focus on the warmth and calm inside my coat and the beauty of the storm outside.

I also worked out. A lot. I would run two to five miles pretty much every morning, and some nights I would return from work so pumped that I'd go for another run. It kept my stress in check and kept me from eating and drinking myself into oblivion, which was my other alternative.

Ramping up the pressure even more was the fact that the glare of publicity had found us. In its day, *Pathfinder* had made the cover of *Time*. During its development, no one seemed interested in making a documentary about the project as it was unfolding. So when *Pathfinder*'s *Sojourner* rover landed, here was a big event—a human story of struggle and success—that went unrecorded, and the documentary guys were kicking themselves.

As a result, when MER got under way, we were swarmed by crews from a half-dozen film companies from the States, Europe, and Canada. It was odd at first, but after a while we got used to it, and like gorillas who learn to have sex while all the primatologists watch, we went on about our business. Occasionally they'd pull one of us aside for a head shot and interview. There were interviewers who asked questions that belied their lack of genuine interest in the subject; those who asked smart, well-researched questions; and clearly some who had their script already written and were simply looking for sound bites to confirm the story line they came in with.

To my mind the most genuine story line was about facing up to the need to rethink everything. As a rule engineers are inclined to dig deeper than the average person out of natural curiosity as well as practical necessity. But sometimes we hold back, usually when we fear the implications of what we might find, which might lead to things we might have to change.

Within MER the degree of awareness and acceptance that the build-to-print mantra was going to have to be violated, and in multiple places, varied widely. Some in project management viewed any discussion of changes in the *Pathfinder* heritage for EDL design with serious trepidation, because it violated the very premise on which the project had been launched.

Greg Davis, the guy Mechanical Engineering team leader Dara Sabahi had asked me to sit beside, was one of these. Greg is a very capable engineer, but his natural bent was to maintain momentum and keep the team focused on our ultimate goal. I came on and immediately began questioning anything and everything, which is a big part of my personal learning process. I questioned our design far more deeply than most of those around me did, followed multiple paths toward what I saw as essential connections, and suggested changes much more profound than anyone else was comfortable with.

Greg did not share my perspective, or at least he saw all too well the trauma that it would inflict on our rate of progress. So he found himself defending the project position, which was status quo, steady as she goes. In his eyes I must have seemed a loose cannon and a threat to keeping us on task. He was afraid we'd fail by missing our deadline. I was afraid we'd fail to achieve the mission objective. This dynamic tension was great at first. We each upheld one side of the debate, and the ensuing discussion was productive for the project. Eventually, though, I think we wore each other out.

Sometimes a project needs the manager who will keep it moving, and sometimes it needs the troublemaker who will stop and question. Dara looked at our situation and ultimately agreed that we needed disruption more than we needed stability. He asked me to take on the title of mechanical systems engineer for EDL, tasked with gluing all the parts together and seeing to it that the sum of all those parts actually worked.

Discounting *Champollion*, this was my real debut as a manager and my first chance to truly influence the larger trajectory of a project. The team was a ragtag band of maybe ten people, and they ranged in age from just-out-of-school Ben, the sweet and sharp, corn-fed Minnesota boy, to Carl Guernsey, an established propulsion expert in his fifties who was nonetheless happy to work in the trenches. There was also John Carson, built like a high school football center and fully equipped with that "keep your head down" tenacity; Denise Hollert, a tall woman who was a "low talker"—almost comically so—which worked wonders in getting everyone to listen; Robin Bruno, whose sarcastic sense of humor only emerged as we became friends; Aaron Fishman, slight and elfin, in his fifties, with a Lebanese accent; and Andrea Kapitanoff, at a self-professed four feet eleven and a quarter inches, tiny, but with a fiery temperament.

I think most people in positions of leadership have some insecurity about their authority, at least at some level. I definitely did, and the way I cemented my authority, at least in my own eyes, was to push to understand my teammates' jobs as well as they did. I tried to win authority through technical credibility. I was able to think of the problems and the solutions to the problems my people were working on right alongside them, which seemed to put them at ease. It also meant that the team meetings had a great deal of intellectual back-and-forth as I encouraged everyone to think about one another's problems and solutions for them. This was great fun for me, and I think it helped the team grow closer to one another and more engaged than they had been before.

Socializing outside of work helped. We instituted something of a Thursday-night bar crawl, called Thirsty Thursdays, which created the bonds that allowed us to be brutally honest with one another at work during the day.

Working so closely together was exhilarating, but I really didn't know what I was doing, and I made tons of mistakes. The most critical of these was in thinking that everyone liked what I liked. That is, I assumed everyone was comfortable with the pushy and intellectually challenging exchange I had first seen in that *Pathfinder* meeting.

I went to every Thirsty Thursday, when really I shouldn't have. Not only would my liver have thanked me, but truth is, a leader doesn't have to be liked by everyone on the team, and probably shouldn't be. There have to be opportunities to bitch and snipe about the person Mother-Goosing your effort—a chance to vilify, wrongly or rightly. I didn't give the team room for that.

What's more, not everyone travels to the beat of the same

drum. Several in the team struggled to find their place, and I think my pushiness may have made them shut down. At the time I was arrogant (okay, even more so than I am now), and I thought the problem was them. They were weak, or they couldn't handle the stress. The luxury of time and some additional perspective and security has taught me differently.

We suffered through it and made it over the finish line together, but I know I made those people less productive than they could have been. A leader has to strike the cultural chord of the team, but you can't make everything a monotone. You have to invite the coexistence of differing styles and do your best to make the culture open to different ways of getting the job done. I still struggle with this a little today, but back then I knew only one tune, and I played it very loudly.

Although the tighter-than-usual teaming may have come from my leadership style, it also reflected, in human terms, the truth of EDL. Because all the functions were so closely integrated with everything else, we had to be in one another's shorts and walk the road together. For many of the mechanical engineering tasks involved in space exploration, there is no demand for tight team interaction, so meetings can be more formal and less frequent. The products and information exchanged among team members can be written down on a list of receivables and deliverables and simply checked off.

But EDL problems demand more integration. The solutions are not as easily understood, and the physics and engineering of those solutions are more intimately coupled. Fluid mechanics, fluid/structure interactions with the parachute, aerothermodynamics with the heat shield, thermodynamics and propulsion and stress with the rocket engine—there's a coupling of functions over a wide range of mechanical hardware for EDL that doesn't exist to the same extent in a lot of other areas of spaceflight. The demands of the task drew the team into a more complete collective understanding of their mutual effort, and really knowing how your component fits into the bigger system is reassuring. It also makes your work all the more meaningful. It invigorates and inspires. Ultimately, it's absolutely essential to getting it right.

Not everyone wants to be a team leader, because not everyone wants the pressure of getting his arms around the whole span and then connecting all the dots. But in my experience, everyone seems happy when he can see those connections and how everything fits together—just so long as someone else is willing to be his spirit guide. If people are encouraged to participate and to try to understand one another's work, the team gets better. When someone doesn't understand some aspect of someone else's component, his questions stem from a beginner's mind, and those questions can be brilliant in taking the team to a deeper level of analysis and problem solving.

"I'm going to try to make you all systems engineers," I told my teammates. This was a team constructed of folks who deliver one part or one element. I was encouraging them to think about engineering in terms of the whole system.

Systems engineering is all about understanding the essential technical behavior and the underlying physics of the spacecraft's functions and how those connections and interaction can be as important or critical as the behavior of any one element. It's understanding the risks of each piece and the emerging behavior of the whole system, knowing what can be known about the system with certainty and what can't, how to balance the risks, and how to make design choices to get a complex system to do what you want it to. Its importance increases to the extent that you are dealing with complicated, coupled, interconnected sequences of behaviors, as we were in EDL.

If the spacecraft is like a wall or building you're trying to construct, then the subsystem bits and pieces—the instruments, structure, parachutes, rocket engines—are the bricks. Systems engineering is both the mortar and the architectural shaping of the wall. The systems engineer has to understand a great deal about each brick, about how the bricks need to be mortared, and about what the wall needs to be.

Up until this point, my work at JPL had been mostly just a job. But now I was getting to fully use my capacity for seeing connections—the technical connections among different elements of the job, but also, and just as important, the basic human connections among the people doing the job. As we enjoyed our Thirsty Thursday pub crawls and held spirited team meetings full of healthy intellectual debate, I watched the team gel more completely than any other group I had ever worked with. I started to realize the connection between the human and the technical. I started to recognize that when you are system-engineering, you are really engineering not only the engineering system but also the human system that creates it.

In our case, the improved intrateam communication and camaraderie pushed us toward a better product. Folks were talking to one another five to ten times a day instead of twice a week; engineers were taking bets on the outcome of tests and really challenging one another to understand the engineering elements as a form of sport. The questions Carl Guernsey, our propulsion guy, asked Ben about his rocket engine mounting bracket made Ben's structural solutions better. Robin Bruno's parachute observations made Andrea's bridle-system work improve. Despite the outsize pressure, everything was getting better, and personally, I was having more fun than ever before.

Chapter 6
TRUTH SEEKERS

AS WE WOUND INTO 2002, WE HAD COME TO KNOW THAT Andrea, the Cog E for the lander bridle system, liked only fancy potato vodka; John, the descent-rate limiter Cog E, bought expensive cars that weren't good for him; and Robin, parachute Cog E/ contract technical manager, would get grouchy without a run every couple of days.

Outside of the mechanical team, I was becoming more at one with the rest of the EDL systems team, led by Wayne Lee, which included Miguel San Martin, in charge of Guidance, Navigation, and Control, and we too were gelling. Miguel and I found a kindred connection and started to really work together to see what needed to change.

All of this meant doom for the status quo.

The closer the systems and mechanical teams got to each other, the closer we got to the underlying truth, and the truth that became obvious to me was that each of our heritage elements was going to have to change.

It was part of my job to take that message upward, but it was not a popular message. We had promised a low-risk, low-cost mission by reusing the heritage items from the successful *Mars Pathfinder* mission. As each of these elements was threatened by change, so was the foundation on which the mission was planned and sold.

The *Pathfinder* landing system we were emulating consisted of the following:

- an aeroshell protected by a heat shield for penetrating the Martian atmosphere;
- a supersonic parachute to slow the spacecraft;
- three solid propellant rockets to slow it further;
- radar to see the ground and to time the rockets and the landing itself;
- a tetrahedron-shaped lander containing the rover, lowered on a 20-meter (66-foot) bridle; and
- air bags surrounding the lander so the bridle can be cut and the lander dropped safely to the surface.

According to the traditional scheme of things, there would be someone working the parachute, someone working the rockets, someone working the air bags, and someone in a leadership position supposedly bringing all these elements together. But in the traditional scheme, that leader was not necessarily informing any one individual about what was going on beyond his or her immediate context, and not motivating them to solve one another's problems with true integration and responsibility.

For MER EDL I saw the mechanical systems lead job as being

responsible for delivering the whole thing glued together in a way that works, with a sharp understanding of both how it works and under what conditions it would fail. This required going beyond the traditional to get everybody deeply engaged in everybody else's business.

Initially Dara had pushed me to reach beyond the narrow and integrate all the pieces. I developed an obsession with drawing everyone together and encouraging folks to help one another understand how what they do individually fits in with what others are doing.

The parachute, for instance, is deployed by a pyrotechnic device triggered by the flight computer. The flight computer is also running Guidance, Navigation, and Control software that is kind of tracking our progress through the sky above Mars. Robin Bruno was the teammate who was working to get the parachute right, but to do that, we had to understand the triggering algorithm well enough to know all the uncertainties that would arise from it. We also had to know the uncertainties of the conditions to which our parachute would be exposed.

Because we needed to understand not just how each component worked on its own but also, and deeply, how everything was supposed to work together, I took on ownership of all the bits, sometimes to the chagrin of the people who actually owned each of the bits in question.

Robin, for instance, bristled under my daily involvement with the parachute and my engagement with the parachute contractor, down to the level of discussing details of seam construction and testing. I felt at the time that it was necessary, but I'm sure that to Robin it felt like micromanagement. It probably was.

Today I look for other ways of taking care of the final, integrated product. I encourage deeper technical penetration by team members if that is what I think we need, and I look to insert myself only as a last resort. In some sense the level of detailed involvement I had over the minutiae on MER could be looked at as a failure of my management. Our backs were against the wall in terms of time, so perhaps that approach was warranted back then. But now I strive to use other techniques if at all possible.

Anyone looking at an organization chart for the project could see that I wasn't a boss of anyone beyond my limited domain of the mechanical system on EDL. I had no direct authority over the teams developing the components we would be gluing together, but within EDL I still saw it as my shared obligation to help find the larger solutions. I tried to model the behavior I wanted to see from our mechanical team members in the larger EDL systems team.

As we confronted this onslaught of unexpected and unwelcome change, the one, inconvenient truth that called the tune was the size of our rovers.

Pathfinder's *Sojourner* had been about as big as a bread box and weighed 25 kilograms (55 pounds). That tiny beast could afford to be small, because it ventured out from a base station that housed the big computer and the big radio for communicating back to Earth. With MER, there was no base station, which meant that all components would be on our rovers, *Spirit* and *Opportunity,* which meant that they were going to be more the size of lawn mowers, weighing in at 173 kilograms (381 pounds) apiece.

We were also trying to increase the science return from MER by having these rovers serve as field geologists. They had stereo vision for surveying the landscape and locating minerals, a grinding tool to cut into rock, thermal-emission spectrometers to analyze the minerals, and microscopes to examine each more closely. The added instruments required a heavier suspension, bigger wheels, steering, radios, an antenna, and a computer, all of which required more power, which required larger solar panels and a bigger heater to keep things warm at night.

Before I came on board, the team had been simply trying to fold these big new rovers into the stock *Pathfinder* landing system. When it became apparent that this wasn't going to work, MER started reaching for help from people who might have more of an architecting, "we need to face the truth and make some changes" bent, which is when Dara Sabahi recruited me.

Not that everyone else wasn't looking for the truth in his or her own way. The whole team had the same objective. But in keeping with Mig's observation about people succeeding through a "constructive interference of personality disorders," I think I was simply a little more driven to find the truth and less concerned about the consequences of getting to it, that is, the need to overturn practically everything. It might have been a bit like my recklessness with my body when I was younger, all those broken bones as a way of looking for some hard sense of reality. I was willing to feel the discomfort of driving for change in a project constructed around things not changing, because it felt real, the kind of real that I had been searching for since my childhood, the kind of real I found when I took that first physics class.

Almost immediately it became clear that to land this much bigger rover using the *Pathfinder* landing system, we were going to need a bigger parachute. The same was true for the rockets we would use to slow the rover after the parachute had done its work and for the air bags to cushion its final impact with the surface. Reusing stuff off the *Pathfinder* shelf had been a lovely idea, but it simply didn't square with the facts.

The bigger parachute was especially problematic, because it would necessarily require a bigger canister to contain it, which would require a bigger aeroshell.

Even with these ripple effects and other troubling implications, I had to make the argument to Richard Cook, the flight systems manager, that we simply couldn't get enough drag out of a parachute the size of the one used in the *Pathfinder* mission. This was a daunting, eye-opening task and a scary change that we could not simply whistle our way past.

Some managers are there because they do a better job at managing. Some are managers because they're just the smartest fucking guy in the room—that was Richard Cook. Mig and I used to compare notes on how, with other supervisors, you could get away with kind of phoning in your management briefs. The guys in charge were not always really paying attention, following your logic, or remembering promises or arguments you'd used weeks or months earlier. That's so not Richard.

Richard is a calm and careful listener. When he's asking a question, you feel almost like you're in the slowly tightening grasp of a boa constrictor. As you move, he moves, and you are forced into intimate proximity with everything you've ever told him, because he will remember all of it. If you ever bullshitted him or exhibited fuzzy thinking, he will gently bring it up. And no matter how deep I was diving in search of the truth, he would always challenge me to go a little deeper. Most of all Richard is exceedingly agile in his thinking. He can recognize a deeper, truer truth and turn on a dime. Mig and I used to say that in a tug of war with Richard, if you convinced him of your reasoning, you would find him running toward you so fast he'd run right by.

For example, we were arguing for doing more parachute testing. This cost more than we had budgeted, so it was not a popular point with most of the managers. Eventually I was making the argument to Richard, and he seemed to be convinced, but then he questioned why my request for the number of additional tests was so small. He said that if my argument was correct, we needed twice as many tests as I had proposed. This was Richard once he had recognized the truth, running way ahead of you.

Every meeting with Richard helped me improve my game, which brought it up everywhere. I started stepping back from my material (PowerPoint slides, usually) and thinking about the essence of what I was trying to say and why I was trying to say it. Then I started thinking of the counterarguments to the position I was advocating and considering whether they were more valid than the support for my point. In short, I started doing what I knew Richard was going to do as soon as I got in the room with him. That made my work better, more balanced, more deeply considered.

Fortunately Richard accepted my arguments more often than not. And as for needing to make the aeroshell larger, we got lucky. We were already using the old *Pathfinder* tooling—a cone-shaped, male mold over which you wrap composite material for the structure. The mold had always extended below the edge of the original spec, so we were able to simply wrap the composite farther down the cone to create what we needed. Other problems, however, were not solved so easily.

Using supersonic parachutes for deceleration in low-density atmospheres is a complex problem that gets into all sorts of strange behaviors, and no other group of engineers faces similar problems. The military people all open their parachutes in Earth's much-higher-density atmosphere, and most frequently at lower speeds. Further, there was no convenient way to definitively test a parachute for our purposes, because testing it correctly on Earth would require an altitude of at least 130,000 feet (24.5 miles), well above what's attainable with the highest balloons or aircraft.

In the 1960s, during the space race with the Russians, NASA experimented with all sorts of oddly shaped parachutes and balloon-like decelerators for planetary exploration. By the 1970s the space agency had settled on the basic disk-gap-band design that was used for *Viking* and has been used for every planetary mission since. The disk is the familiar rounded cap with a vent in the peak that helps the chute open correctly and fly straight. The gap is just that—an open section below the disk that provides further "geometric porosity," or venting for stability—linked by suspension lines to the band, which constitutes the "bottom edge" of the parachute.

There was certainly room for disk fabric that was thicker and therefore stronger, and there was always room for a better design— a way of varying the disk or the band—that might increase performance, which is to say increasing drag while maintaining stability, without adding bulk.

But even after we grew the size of the canister that would contain the parachute on its way to Mars, which required making the whole shape of our vehicle a little different from that of the one *Pathfinder* flew, we still needed more chute. And between the need to redesign not only the chute but the canister, it was June 2002 before we were ready to test.

Our parachute vendor had access to a National Guard gunnery range in Boise, Idaho, as well as to some heavy-lift helicopters. So we took our new, larger design to the range, attached one of the chutes to a huge dart about the size of a cruise missile—and about as heavy as a fully loaded, full-size pickup truck—then took it up almost a mile into the air and dropped it. It wasn't the definitive

130,000-foot supersonic test, but it was the best we could do. The parachute opened and exploded, fluttering to the ground like a dead bird. We were stunned. "What the fuck just happened?"

Had we overtested with our procedure? Was the fabric being overstressed?

We tried it again and boom—the same thing.

We took an evening to think about what had happened, double-checked our calculations on the fabric stresses and the test procedure, then tried one more time. More shredded orange and white parachute fabric fluttering to the ground.

We were launching in exactly one year, and the parachute was one of the elements that was supposed to have been wrapped up long before now. All the bits were supposed to have been completed, with nothing left to be done but to bolt them together, but now we were staring at a complete redesign of our already redesigned chute. We were not just behind schedule. We had no idea what to do, because we had no idea what was wrong.

I told Dara about the troubles we were having but asked him to keep it quiet. There were so many uncertainties in dropping things from a helicopter that we were going to test again the next day. If the failure was just a test artifact, I didn't want everyone to get worked up unnecessarily.

But the next day it failed again. So we called in emergency triage in the persons of Juan Cruz, an aerodynamics specialist, and Prasun Desai, a trajectory simulation specialist, both from Langley Research Center in Virginia, to help us run a trajectory simulation in Boise. They arrived sixteen hours later.

Our team, beefed up with these troubleshooters, took over a conference room in the Holiday Inn and littered it with laptops, papers, and blueprints of parachutes. To have any chance of finding a solution, we needed to know the exact conditions under which our chute had failed.

Falling from the helicopter, the dart started out slow, then accelerated, and then the chute tried to open. But how fast was it going at the point of failure? This was critical information that would be exceedingly hard to obtain by standing out in an open field looking up at the sky.

After we tested again and failed again, things got a tad stressful. My normal algorithm of food or drink or exercise became one of food and drink and exercise. After our epic sessions in the conference room, when we had absolutely nothing left in our brains to work with, we would retire to one of several Boise nightspots and slam back the bourbon. The next morning we'd have to start extra early, making time to run the trails in the foothills of the Sawtooth Range just to shake off the previous day and be ready to face a new one. It wore on all of us.

Eventually Richard called and said, "Guys, stand down. You're beating yourself to death, and you still don't know what's going on? Take some time."

We took his advice, and then we reached out to every parachute expert we could find for consultation. Our parachute contractor, Pioneer Aerospace, gamely allowed us to include experts from their industrial competitors in that number. This team of consultants included folks who had worked on weapons systems from Sandia National Laboratories and others. We talked on the phone and tried to make sense of it. Getting nowhere, we made a date to all converge on Pioneer's facility in Connecticut in two weeks.

As it happened, a day before we were to meet, the culprit was unearthed. Al Witkowski, from Pioneer, had been staring at the drawings for the parachute when a bulb lit up. When we got to

Connecticut, he showed us. As we stared at the drawings and thought about the shape the chute takes when it inflates, the same bulb lit up all around the room.

In this case, we had looked at a drawing of the parachute in its shape after it had been sewn together but before it had been exposed to aerodynamic loading. Unfortunately we had chosen to build the structure to handle the loads placed on the chute based on the wrong model—the uninflated shape. In reality, when the parachute fills with air, it takes on a very different shape, and in that shape some of the loading is in the direction opposite that which we had built the structure to handle. When the parachutes filled with air, they simply came apart.

The chute is made out of polyester and nylon and forty-eight Kevlar suspension lines. You sew the lines to the fabric, but you never want to put that seam in tension, which would ask the chute to hold together entirely by the strength of the sewing thread. You put the lines on the outside of the fabric so the fabric wants to push itself into those lines, with the thread simply holding it in place. But our parachute formed enough of an orb that our lines wanted to pull *away* from the outside. Since one never gets to see a parachute inflated up close, none of us had recognized that.

By now it was obvious that we were not going to get everything that we wanted in our redesign. We couldn't make our chute any bulkier, and the trade-off was performance, in slowing us down, versus strength.

Given the limited time, we couldn't afford to do a series of designs to find the one that had the best drag at acceptable strength, so our only choice now was to go parallel. So we set out to make five different redesigns and then test them in quick succession under controlled conditions.

In doing this, we were dooming four of our five design and fabrication teams to fruitless labor, because only one path was going to be followed to completion. We had to be creative enough to come up with the five configurations, we had to have enough human resources to man those teams, and we had to be skillful enough to motivate everyone to give their all, knowing that they had a four-out-of-five chance of being a sacrificial lamb.

This is where the particular talents of Wayne Lee, the chief engineer for EDL, came into play. Wayne was a big sports fan, and he was like the coach who can motivate players to give their all and to feel good even when they're about to throw up. He was not as aggressive about finding the hard truth in all technical matters, but he had the great virtue of listening, and he could be persuaded to do what was right any time someone dug down to the truth and pointed it out.

In this case, Wayne was the one who came up with the key idea of going to a wind tunnel—not just any wind tunnel, but a gigantic wind tunnel, in fact, the world's biggest: the National Full-Scale Aerodynamics Complex at NASA's Ames Research Center, near San Jose, California. The rub was that, to open a chute in a tunnel, you first have to fire it out of a mortar. "Nobody's going to let you fire a mortar in there!" we all said. What we didn't know was that Wayne's father-in-law was a former director of Ames.

Under the arrangement that Wayne worked out, we would be able to do simulated drops for a solid week, all day every day, under very precise and easily replicable conditions.

We took our five new designs into the tunnel, spooled the wind up to around 120 to 135 miles an hour, and fired our chutes from the mortar. The test conditions were perfect, and the facility was a dream. The same was not true of the parachutes. Some of them failed to fully inflate; instead they partially inflated, dancing in and out in a strange, swimming pattern known as squidding, a rare and unwelcome behavior of poorly designed parachutes. NASA had worked with parachutes like this for thirty years and never seen such a thing. It was as if the chute had a personality disorder and just wanted to fuck with our heads.

The extraordinary thing about a wind tunnel is that you can walk around in it and closely observe phenomena, from all sides and for extended periods, that you'd otherwise have to be falling from the sky in order to see. As we examined the squidding parachute, Robin Bruno, the parachute Cog E, and I kept coming back to the vent. We said, "That hole's too big."

The manager from Pioneer said, "No way." He was fairly adamant about it. We wanted to measure the vent, but he insisted that there was no need; the vent was exactly what it was supposed to be.

Fortunately Pioneer had also brought along Roy Fox, one of the world's leading parachutists and an experienced elder statesman of the parachute community. As a consultant rather than an employee, Roy was above the fray, and I asked him to do me a favor.

While we distracted the manager, Roy measured the vent. It turned out that this opening in the peak was indeed the culprit.

When we'd asked for our array of designs, apparently we'd made an error in transposing to scale. Ideally you want to keep the relationship between the size of the vent and the size of the disk constant. *Pathfinder* had used a version of the *Viking* parachute that had doubled the band. We were looking for solutions that were somewhere between *Viking* and *Pathfinder*. The way we were generating those intermediate configurations was to start with the *Viking* configuration and grow toward the *Pathfinder*'s, but as the band got deeper, the design team also put more area into the vent, which meant that the design was on its way to becoming a windsock.

So now we knew what was going wrong in the wind tunnel, but we had run out of time to make it right. We were less than a year from launch, and while we'd been scratching our heads in the wind tunnel at Ames, most of the spacecraft had already been packed up and shipped from California to the launch site at Cape Canaveral.

We'd been looking for an optimum solution, one that got us the most bang for the buck, and in the end we simply ran out of time. So now I had to brief Richard on the idea of using the *Viking* parachute configuration, which would produce more drag with the existing size but have less stability. Which is when he recalled in great detail the conversation we'd had six months previously. "When you told me we couldn't get more drag out of the parachute, you weren't entirely accurate. . . . This *Viking* chute was available."

But he didn't hold my backing and filling against me. Ultimately he said, "Just fucking fly something we know we can build." So we crammed the biggest *Pathfinder*-style parachute we could into the oversize parachute mortar in the oversize aeroshell and called it a day.

One of the problems with space exploration is that we never have enough iterations to allow us to fully learn from our mistakes. When it takes five years to launch and then two more to realize success or failure, there's a time constant that defeats learning. By the time the deepest truths emerge, most of the humans who were in on the creation very likely have moved on. If we hire a contractor that saves us millions but in the end their work fails, it's hard to know what caused the failure, because we don't do enough missions to gather the right kind of statistics, so we're left with too much supposition.

In some senses what we do when we develop a one-off space-flight system every five to ten years is akin to a modern high-tech start-up. It takes a few years, every time it's different, and you don't get a lot of practice.

All of this is why asking the right questions in the first place, and then listening deeply to the answers, is vital to embarking in the right direction.

What are the right questions to ask? What do they look like? What do the answers sound like?

Well, to start, the right questions are those that are essential and profound, which means those that are architectural in their dimension. Frequently the right questions define the watershed between differing strategies of solutions or crisply define an essential risk.

In the case of our redesign of the *Pathfinder* EDL system to accommodate a much bigger rover, the essential question was "What must we do to accommodate this bigger rover?"

I was in such a breakneck effort to get the message out about the need to make the necessary changes that I did not really stop to listen to the full answer to the question. I got the part about more parachute drag, but like Christopher Walken in the *Saturday Night Live* skit, I kept demanding "more cowbell." I wasn't listening to the full and balanced end of the answer.

Over time I have found that most of what gets in the way of my seeing the universe as it really is comes from me. Excitement, pride, preconceptions of what the universe should look like all get in my way, but the most common block is fear. Fear that I won't measure up, fear that I can't get the answer to the open questions, fear that I will let others down, fear that I will shrink in the eyes of others and in my own esteem—all of these can sit in between me and understanding. I don't think I'm alone.

When I scribbled "Hold on to the doubt" on those cheat sheets back at College of Marin, I was encouraging myself to breathe and let go of the fear so that I could sit beside the open question and really understand it. In a state of calm and with quiet contemplation, I could best answer the academic questions. It turns out that the same is true for real problems of engineering, and innovation, and even leadership and personnel management.

With the MER parachute development, I had in some measure failed to hold on to the doubt. I had helped ask the right questions with the team, and we had part of the right answer in that we certainly needed more drag and therefore a bigger parachute, but I had let fear of not getting enough drag pull me away from the question of how much drag was enough.

Through all of this, and following Dara's lead, I tried to help people feel comfortable in the sweet spot where passion and curiosity, combined with common sense, could operate safely.

While Greg had held meetings once or twice a month, I had meetings with the team twice a week. I was driven not only by the need to find the bedrock truth but by an addictive pleasure in the collective effort of getting down to it. I think in those team meetings I was channeling Dr. Prata, my first physics professor. He had always welcomed with infectious glee the prospect of connecting the dots, and as I connected the dots with my teammates, I let my glee show. Human curiosity is a powerful feeling. If I was lucky, I was making a space that was safe for the team's collective curiosity and urging them to come play with me.

In fact, it is our curiosity that we need to engage and our fear that we need to hold at bay, or release, if we're going to find what we're looking for. It is as if there are two forms of decision making: fear based and curiosity based. In fear-based decision making, we find ourselves wanting the answer as fast as possible. We don't really pay attention to, or care, what the answer is—we just want something, because the open question makes us anxious and fearful. In curiosity-based decision making, we use one of the core traits of our species to pull apart, examine, and wade into the open question. In my experience curiosity-based decision making yields much better solutions.

When you work in any moderately large organization, you learn the classic styles of managing up, managing down, and managing to the side. In an environment like JPL's, which incorporates multiple lines of reporting, there's an advantage to going off-road and finding less rigidly geometric ways of exercising influence and maintaining alliances to get things done. So I went around collecting people who I knew saw the world the same way I did, exercising the art of seduction like a politician trying to get his legislation passed.

What I found to be most honest and most effective was to have all the groups I interacted with good-naturedly bashing one another. The goal was to find a way for ideas to win rather than people, so it was important to set up this jousting as a test of competing truths rather than competing egos or hierarchies.

To get to that place where the conflict could be creative rather than combustible, we also needed the team to remain very close personally, so I saw part of my job as being a kind of high-tech camp counselor.

Growing up in Marin in the seventies, I learned from my parents early on to be open minded and accepting, but now I realized the need to take it further. I wanted to genuinely like the people I spent my days with, if for no other reason than that it makes life more enjoyable.

We work with our teams for eight to twelve hours a day, five days a week—a huge chunk of our waking life. So I set out to try to find one thing to love about everyone. Sometimes you can't get all the way to loving the whole person, but maybe you can love the goofy way they show up in a bike helmet, or the way they have those cat pictures all over their work space, or dress fastidiously, or love to organize thoughts on spreadsheets, or how they can't help dreaming of solutions that appear off topic. Oddly enough, in time the love actually happens.

In the same way, your frustration with someone doesn't have to be with him or her fundamentally as a person. You try to limit your frustration to just the behavior, because getting to the hard truths requires people who can show affection and respect for one another at the same time that they feel at ease showing complete disrespect for ideas that just aren't going to cut it.

In the years to come, I would sponsor "EDL Lunches," with five to twenty-five members of the team, going off-lab to a place called Nicole's, in South Pasadena. Sometimes we'd talk about business, but mostly we'd just try to be human, talk movies we liked, restaurants, our opinions of our children's schools, or awesome wackiness found on YouTube. Eventually I placed a calendar invite on the EDL schedule called Lunch with Enrico, to protect that lunch hour for team members to gather for fellowship.

The "Enrico" part is an homage to the physicist Enrico Fermi, who was a master of this kind of family feeling. There's a story about him on the day they were about to create the first sustained nuclear fission reaction. This was at his lab beneath the University of Chicago football stadium in Hyde Park. Three rods needed to be pulled out to set the reaction in motion. In some sense Fermi's entire career had been leading up to this one huge moment. But on the big day, just as the technician pulled out the first rod, Fermi noticed the clock. "Stop," he said. "Time for lunch." So, as usual, his team gathered around a big table to enjoy one another's company, sharing stories about family and whatever else was going on in their lives.

We're social animals, and our brains work better in an environment where we're committed to one another and enjoy one another's company. That's the only environment in which it's safe enough for the small voice of real truth in each of us to clash with one another fearlessly and energetically.

Following Dara's example (but toning it down a notch), I developed a way of mixing the rational and the irrational right up to the edge of looking a little bit crazy. (Or is that just the excuse I use to rationalize my natural behavior?) I saw it then as like a gazelle that dances in front of a lion to throw it off balance, or the mime or the court jester who plays the fool. In retrospect, I sometimes call it dancing wrong in front of the lair of right to coax it out. The crazy part is being willing to be wrong in search of right and not worry about looking foolish.

Most people are always a little worried about being wrong, but that's where my youthful tendency to fling myself off rooftops and walk atop fencelines has persisted—I'm just not concerned about being cautious. And if you demonstrate that you're willing to be wrong in pursuit of finding right—and freely admit it when you are wrong—everyone else relaxes and feels free to put themselves out there in that same search.

I remember challenging Rich Hund at Lockheed Martin in Denver, the contractor in charge of our aeroshell. We were debating the use of the seal between the heat shield and the back shell. We were in Lockheed's facility in the foothills of the Rocky Mountains, and I bet him a beer that he couldn't prove the structural need for the seal; that is, the need for the seal to act as a load path. Though I secretly thought he might be right, I was willing to take the wrong side of the debate to prove it to all of us. Rich had his guys dig deep into the analysis, and before the day was over, I was buying beers in an old saloon in a tiny town up the mountain from the plant. The beer tasted especially good thanks to the camaraderie we shared as we drank it and was a talisman for the understanding we had mutually established.

The key to searching for the truth is to hold passionately to your beliefs while simultaneously not feeling entrenched in your position, to be able to let go of the need to defend it in order to save face. It's almost a Buddhist thing, where you're not necessarily free of ego and concerns about status, but you're able to sit with them and maintain some objective separation. It's about letting ideas win, not people. It's about finding what's right, not being right.

There was another way my mental process changed as I developed as a leader, and once again it was inspired by Dara. He often displayed a very un-engineering approach to engineering problems, and he encouraged me to do the same.

Most of an engineer's academic training focuses not on problem solving but on problem definition. I would bring Dara a problem, fully defined, and I would start to describe in detail the various and, to me, somewhat beautiful folds and corners of the problem definition, and Dara's eyes would glaze over. He didn't care. He had heard enough to recognize that we had a real problem, and he was already searching his brain for a solution. Getting used to this, and ultimately embracing this approach, required me to resolve (or maybe embrace) some seeming contradictions.

Holding on to the doubt means you listen to the problem until the deepest truth presents itself. But that, of course, raises the question: How do you know when you've gotten down to the deepest truth?

This is especially problematic in engineering, where you may have seventeen channels of measurement data giving you hard facts, but there might be five thousand other channels you could measure just as well.

Before the Iraq War, Secretary of Defense Donald Rumsfeld famously spoke of the different bins into which all states of knowledge can be cast. The hard facts are what he called the known knowns. We have plenty of these in engineering: the acceleration of gravity, the number of protons in a helium atom. There are known unknowns, the things we know that we don't know, like the exact calorie count of that burger you are eating or the wind speed and direction on the day you plan to land on Mars. The two other bins get tricky. The bin of unknown knowns, which Rumsfeld has used to refer to things you think you know but you do not. This is what I like to consider self-delusion. A cautious and honest mind might be lucky to largely avoid unknown knowns through self-questioning and asking others to set their eyes on your work. The final set, and perhaps the most important, is the unknown unknowns, the things you don't even know that you don't know. Intuition, which taps lightly into the irrational and, I believe, integrates a larger set of the rational than is available to our conscious mind, is a way of bringing in the unknown unknowns and of creating a broader model. Sometimes that model is so broad that you can't articulate it. Sometimes it can't be integrated at a conscious level. It appears to us as a "feeling."

For however much his passion may have overstepped the bounds, Dara always seemed to know immediately and intuitively the course he wanted to take. He had supreme confidence, and experience had shown that his confidence was justified.

Most of engineering is explicit—it's all about the math, the known knowns—but there's also the "inner voice" of engineering judgment, your confidence in what you know to be true. Then there's your Spidey sense, the confidence in your gut that you learn to trust beyond the horizon of the hard data. It's based on the data you have, but it also encompasses confident extrapolations. As such, Spidey sense is the essential counterbalance to the rational minimalists who deconstruct and analyze every argument into its elements and miss the intuitive altogether. It is the part of our awareness that is reaching out to get a grip on the unknown unknowns.

Dara brought back in the intuitive and instinctive thought that can help bring in the missing pieces of the whole analysis. In this he was more like an artist, or the old-world tribal patriarch whom he sometimes appeared to be channeling. Dara trusted his Spidey sense to the extent of taking drastic action based on it, as when he put me in charge of comet stabbing on *Champollion*. Essentially what Dara had mastered, and what he was pushing me to master, was the integration of the subconscious and conscious minds.

Our screwups today are frequently errors of omission. Science has been pretty good at discarding the fictions we once used—the four humors, the music of the spheres—to fill in the physical laws we didn't understand. So when we settle for self-delusion these days, it's usually because we've stopped short in our truth seeking and have accepted a subset of the universe in our model as being the whole universe. In engineering we might omit a term in an equation because we don't think it is important—too small to matter. More frequently we omit a test or an analysis from our plans to save money and because we think we can get by without it. These are willful omissions, and although they can sometimes wreak havoc, they are never as truly dangerous as the unconscious omissions, the failures of imagination, the true unknown unknowns that have not been considered. For we do not ever debate the things we have failed to recognize as being important. Our Spidey sense tries to subconsciously model those unknown unknowns in some fashion, accounting, perhaps, for the sheer possible magnitude of what we don't consciously know that we don't know.

Some of my teammates used to tease me about my appreciation for Spidey sense. They compared it to Stephen Colbert's "truthiness." And, fair enough, "trusting your gut" can be a blanket excuse for choosing convenient thoughts or ideas over the truth just because it feels right. This path of self-delusion is the greatest threat when you try to include the subconscious mind in your decision making.

But the subconscious is also very, very powerful. Each of us has in our personal experience a sudden realization of the perfectness of a solution or the brilliance of an idea. We make those recognitions without having explicitly gone through "the math" in our heads—without walking through each step of the logic. We do it because we have subconsciously worked it out, in a millisecond.

So how do you harness the subconscious, your Spidey sense, without running aground on the shoals of self-delusion?

You talk about it. You share your thoughts and compare your Spidey senses with those of your teammates, external advisers, and nonadvocate reviewers. Through that process one frequently mines the subconscious and drags the ore into the sunlight of conscious consideration. It allows the great debate to occur.

All that said, the great debate does not always answer the

question. Sometimes there isn't time for true conscious consideration of all the possibilities. Sometimes a spacecraft is hurtling toward Mars and you have a few hours to choose whether to change the center of navigation or not. In those moments you have to check with yourself, do your best to make sure you're not acting out of convenience, and then go with your Spidey sense.

Chapter 7
THE DARK ROOM

WHEN YOU ARE SOLVING A PROBLEM OR INVENTING SOME-
thing new, or even reinventing something old, you can find yourself
stuck, with no solution appearing to be viable. It is a frustrating and
terrifying process, especially if you are on a time line with an un-
movable product debut date, set by celestial mechanics. That place
is what I call the Dark Room.

While working on the Mars Exploration Rover mission EDL,
we had faced the fact that we had to change the parachute. It had
been a difficult realization for the team, but we had accepted it. The
problem was that the changes were not going to stop there. Even
after all that grief, we still needed further deceleration and guidance to get to the landing site, which was the job of retro-rockets.

The rockets for *Pathfinder* had been custom-made. Nobody
wanted to hear about an expensive redesign, but that was yet again
one of the painful truths we had to face to safely transport these
much bigger rovers to the surface.

The *Pathfinder* rockets had been 5 inches in diameter and about 3 feet long. To accommodate our heavier payload, we asked our vendor, Morton Thiokol, in Elkton, Maryland, to figure out how to make a rocket that could produce three times the thrust, which led to rockets that were 10 inches in diameter and 2.5 feet long.

When the Army drops really big tanks out of a C-130, they sometimes employ a retro-rocket between the tank and the parachute to provide a final deceleration just as the tank approaches the ground. They hang a plumb bob maybe 30 feet below the tank, which hits the surface first, then activates the rockets to fire, offsetting the gravitational pull just before impact.

Pathfinder had stolen that idea and used a similar system, only it relied on radar instead of a plumb bob to trigger the rockets. But radar never gives you 100 percent accuracy, especially not when it comes to measuring speed. And while a tank landing on Earth might be coming down at a rate of 25 miles per hour and you want to reduce that to 5, on Mars the rate of descent is 150 miles per hour, so fast that you have to make the decision to fire the rockets from a quarter mile up. This not only makes the job a lot harder on Mars; it also adds a much greater degree of uncertainty.

Unfortunately that's not the worst of it. On Mars, as the parachute plummets at around 150 miles per hour toward the surface, winds can make it swing from side to side like a pendulum, all the more so when it has a rover hanging below it on a bridle 20 meters (66 feet) long. Fire the rockets while you're on a wide swing and, while your objective may be to slow your vertical descent, you will add substantial speed to your horizontal drift. The net result is that the air bags can hit the surface in a shearing, glancing type of blow that can tear them to pieces.

Miguel had the idea of creating additional rockets to steer should the main rockets need to be fired on one of those wide swings. The system he dreamed up incorporated a spare camera patched into a camera line already on the rover to help determine horizontal speed. The camera could look down at the apparent movement of objects like craters on the ground, take multiple images, then feed the data into an onboard computer that could calculate our speed and decide when and/or how to deploy the steering rockets. It became my job to make the mechanical engineering of Mig's plan work, by developing those rockets and figuring out a way to get them into the back shell. My former student Ben Thoma and the slight-of-build Aaron Fishman worked slavishly to make that system come together. We would call the system TIRS (transverse impulse rocket system) for the agony it put the team through with this late development!

The machine vision algorithms needed to take three images of the ground, find objects to track within them, and calculate ground speed were another matter. The job of developing these algorithms fell to Andrew Johnson. His algorithms could look at images, find and identify objects (craters, mountains, and so on), find the same objects in the next image, do the rectification between one picture and the next, put them in the same reference frame, and calculate the speed by comparing the amount by which they've shifted. We called this system DIMES (descent image motion estimation system).

Because I was part of Wayne Lee's team helping facilitate and integrate multiple aspects of EDL and not just those in my mechanical engineering domain, I was a part of broader discussions with other teams, where I often found myself passing along Mig's ideas.

Miguel and I share a need to always strip down the thing to its essence. What is the key element of the solution? What is the most important part of the particular problem we are facing?

My adviser at Caltech was a fabulous Scotsman by the name of

Ron Scott. He used to say, "Your brain is like a sausage casing. It can only hold so much, and if you try to stuff in too much, something important might get pushed out, like your daughter's birthday."

Mig and I share an attraction to the essence, a need to reduce mental clutter. Because of this, we tend to like the same solutions, fear the same problems, and see things in similar veins. I always like talking about Mig's solutions, because they are almost always elegant, simple, and essential.

It was my job to get the mechanical engineering for EDL integrated, and it was my inclination to think at a systems level. That's why I was speaking about TIRS and DIMES at some system-level meeting and feeling overjoyed to play in this rich intellectual field. Reaching across the boundaries of the subsystems was playing in the grass on the other side of the fence.

The next day I was walking down the street with Miguel, telling him about that meeting. He stopped and seemed to muster something uncomfortable from deep within. "Adam, I should be in those conversations."

It was a simple statement, and also undeniably true, but it caught me off guard and made me feel guilty.

I said, "You're right," but in a casual way. Meaning, I wish someone had asked you, but since you weren't there, I passed on the information.

He said, "No, I can't do my job if I'm not in those conversations."

He had put his finger on the nut of it. I may be swept up in the joy of playing with his solutions, but I'm not Mig. He needed to be playing there, in the grass on the other side of the fence, with me. In my enthusiasm I had overstepped, and also overshadowed my friend. I was so gleeful about being in the big show, and using more of myself, and being in a position to materially fix some things, that I'd lost sight of myself in relation to my closest colleagues. I was keeping him from growing, and blocking his full potential contribution to the design.

The lesson for me was one of temperance and moderation—not an easy one for me, and one I still work on today.

The joy my work brings me can sometimes be blinding. As with many people, or perhaps all of us, my greatest strengths, when concentrated too heavily, become my greatest weaknesses. Using your full self can be a great thing, but not when it crowds out other people.

The drive to invent TIRS and DIMES came from the need to protect our air bags from being destroyed on impact. After deployment of the parachute and the rockets, the air bags were going to be our ultimate deceleration device. And as was the case with every other element of MER, our initial plan was to use the same heritage design that had been developed for *Pathfinder*.

Those most knowledgeable about the *Pathfinder* bags said that the old designs would be able to handle the extra weight. But memory is too kind. We often see things in the past as better than they were.

We needed to do our own testing. Unfortunately the air bags we used on *Pathfinder* were not an off-the-shelf item. Even the fabric used was custom spun and woven, so just coming up with the materials might take six to nine months. Fortunately, not long after the *Pathfinder* landing in 1997, we donated some of the spare air bags to the Smithsonian National Air and Space Museum. They had spent the intervening three years either sitting on display or boxed up, as needed. We called the museum and said, essentially, Can we have our air bags back? Happily, they were willing to pull them out of storage and let us try them out.

The job of an air bag that's landing on Mars is to hit the surface at about 20 to 40 miles per hour; to bounce maybe ten or twenty times, cushioning the impact and protecting the spacecraft inside; and, most important, to not puncture. But because an air bag operates on the basis of absolute pressure rather than gauge pressure, and the Martian atmosphere is less than 1 percent of Earth's, if we were going to test at all, we needed to test in near-vacuum conditions. But because each air bag is the size of a small conference room, to do that, we were going to need a really big vacuum chamber.

We took our bags to the world's largest: the Space Power Facility at Plum Brook Station, part of NASA's Glenn Research Center, in Sandusky, Ohio. The chamber—122 feet high, 100 feet in diameter, and carefully suspended inside a nuclear reactor containment facility—was built in the sixties to test nuclear power stations for use on the moon. The walls were made of concrete 6 feet thick and lined with 1 foot of steel. It was a facility to satisfy even the maddest mad scientist on a gargantuan scale.

Dara and I arrived at the test facility after dark on a winter night after driving through whiteout conditions on unmarked roads through a 6,400-acre wilderness.

Tom Rivellini, the air bag expert from *Pathfinder,* was already there when we arrived, as was Mig. Mig is not a mechanical engineer, but in JPL's open culture, we welcome as much cross-pollination as we can bring to bear, and Mig is always a good bee. These are also really cool tests to watch.

Inside this vast, empty space, we set up cameras everywhere to capture whatever happened. Back on *Pathfinder,* to create acceleration, Rivellini had invented a huge slingshot that could fling air bags onto hunks of rock affixed to a ramp that could be positioned at different angles. We reestablished this test environment, setting out an array of rocks as analogous as possible to dangerous rocks thought to exist on Mars.

As we slammed the first bags—spares we'd kept from the *Pathfinder* mission—onto our rocks at 30 miles an hour, we watched in slow-motion video as the rocks blew open holes big enough to walk through. This was not good.

We then took the set we'd retrieved from the Smithsonian, the last we had, and flung them onto the rocks. They blew apart as well.

"They'd been sitting around too long, perhaps under fluorescent lights," Rivellini and Dara said. "The material could have degraded."

But was this answer too easy? I suspected a rosy (but inaccurate) glow might be coloring their memory of how well the air bags had performed on *Pathfinder* and even of how thoroughly they'd been tested.

We struggled with the bags through several design iterations and conducted some puzzling trials. One rock in particular, which we came to refer to as the Black Rock, became our nemesis. A smooth and unremarkable stone the shape of a cow's liver, with a light ridge running along its top, it didn't look that dangerous. It didn't damage our abrasion layers, but it reached in and caused the bladder inside the bags to rupture. We made molds of the Black Rock so we could replicate it and put it on other parts of the ramp and slam the air bags into it. What we realized through these tests was that we'd landed *Pathfinder* on Mars with a system that would have been defeated by the Black Rock, and perhaps many others. So maybe we'd just gotten lucky. We couldn't and wouldn't count on that luck holding for MER.

The inside surface of an air bag is coated with silicone to stop the gas from coming out. That layer—the bladder—takes all the tension of the pressure of the gas. Then four abrasion layers of gauzier Vectran (a cousin of Kevlar) are put on top of that, on the outside, to take the bulk of the abuse. In some ways this is like a modern tubeless automobile tire. The tread provides grip and protects against abrasion, but it is the belted material—nylon or steel fibers—under the tread that keeps the pressure in. On our bags all that armor was getting torn through, so we went to six abrasion layers, but even that got torn. We tried abrasion layers of alternating thickness, but they got shredded. All the time the mysterious Black Rock was magically reaching through the abrasion layers without doing much damage to them and busting the bladder.

There comes a moment in every creative or innovative process when you're not only lost; you're not even sure where to find a map. Spending time in this Dark Room is terrifying, but there is no easy way out. You have to stay calm, hold on to the doubt, listen to the problem, and keep thinking of solutions while avoiding the mind-locking panic that you won't find one in time. We were in the Dark Room with those parachutes, and our only way out was persistence. What solution could we use that could survive the abrasion? How could we beat the Black Rock?

After almost a year of groping in the darkness, we found a ray of light, a spark of an idea, a pathway out. It was a rather arcane solution, and hardly new, an idea that had been around since the first pneumatic bicycle tires in the nineteenth century: the inner tube. We needed a second bladder, held inside the first but not under tension of any sort. All we needed this inner bladder to do was to keep the gases in, so it could be cut larger, meaning that it would have slack, and these unstressed fibers would be infinitely more resilient.

The outer bladder would still be there to absorb the stresses coming from the outside, and thus it would be taut. We would build the outer layers just as we had been, and if they were bested by a rock strike, they would fail, but the air bag would not. The inner bladder would not rupture but move out of the way during the impact; then it would bulge out a bit, but it would keep the precious gases in. As long as you didn't get two hits in the exact same place, as long as lightning didn't strike twice, the air bag should survive. In test after test, this new approach (though very old in some ways) proved its value. This new design was just plain tough.

Even with this problem solved, the roller coaster that was MER just didn't want to stop. There was one more element standing between us and a safe descent, and that, too, had us circling in the Dark Room for a long, long while.

When you're on the parachute, before you inflate the air bag, you lower the lander below the back shell. You have to do this to get the air bags away from the rockets and to make the whole system stable when the rockets fire. To get this lowering effort right, you need to manage the rate of descent on the bridle.

For *Pathfinder* we'd used a descent-rate limiter developed from a pilot escape mechanism for the 747. It looks like a retractable dog leash, only with a stainless steel tape coiled around the spool instead of a nylon leash and a much more substantial centripetal brake. The pilot pops open the cockpit window, clips the device to the cockpit, and rappels down to the tarmac on the leash.

To slow our lander below the back shell, we tried to use one of these descent-rate limiters right off the shelf, but the MER lander (1,190 pounds) was much heavier than the *Pathfinder* lander (800 pounds), and the steel tape kept breaking. We went through conniptions trying to find a solution.

The toughest thing about the Dark Room is that you don't know if it's the darkness before dawn or the darkness before pitch-black. Your stress hormones are flowing, so there's no shortage of motivation to stick with it. The challenge is to avoid a sense of hopelessness.

These days I can usually buck up my team by sharing my experiences of being in the room many times before and always managing to find a way out, but back then I was just plain terrified. I can remember telling a friend, "If we can't figure out this problem, we don't have a way to Mars!"

This is the point at which Tom Rivellini reminded me of the old saying, "The coward dies a thousand deaths." There's plenty of time to confront "the worst" if it comes to that, so there's no point rushing ahead to assume that you've reached it. Today I tell my teams, "Just keep working, and if death comes to visit us, let us be surprised." Where there is will and ingenuity, there has always been a way.

And there is a virtue in staying in the Dark Room until you completely surrender—not give up but completely surrender the ego, which means that you stop trying to force the existing solution. That's when a breakthrough can occur. When you're in that neutral zone and no longer striving, when you've reached that ultimate humility of questioning whether you even have the right to call yourself an engineer, the subconscious is free to kick in. And then, when the new solution presents itself, you're so completely drained and empty that you're wide open to it, devoid of all prior notions, completely receptive. It's the perfect blend of the conscious and subconscious working together, engineering judgment and Spidey sense.

I was trying to egghead my way out of the room by doing all sorts of fancy stress analysis to try to figure out why the tape was breaking. Instead Tom Rivellini came in from left field to design us out of it. I was trying to rely on the thing we had this metal tape wound around, the centripetal brake, and to figure out how to treat the opening of the device to keep the tape alive. "Forget the tape, forget the spool," Tom said. "Just use the brake and something else for the line."

Instead of the steel tape, he suggested a Kevlar line, along with a device that was more of a capstan winch than a spool. With this approach the line was not stored on the spool at all but simply wrapped around the winch a couple of times, like the jib sheet on a sailboat, and the winch was attached to a brake, so the speed never changed. It was at once a bit more complicated and a bit less functional (we liked the changing speeds that happened when you wrapped the tape around the spool like a yo-yo), but it worked, and it worked well.

We emerged from the Dark Room triumphant, our system complete. Our parachute issues had been put to rest, our air bags were squared away along with our rockets, and now we had our descent-rate limiter.

By the beginning of 2003, all the parts had been built and were being assembled, tested, then disassembled, sent to Cape Canaveral, then put back together and tested again.

This is the point when mechanical engineers either go with the team that bolts it together, the ATLO team—Assembly, Test, Launch Operations—or move on to something else. My skill set is more design architecture—system-level stuff—so I moved on to something else: a new project called *Mars Science Laboratory*. In this I was looking for an entirely different kind of challenge: to take the lead in designing and building a new class of rover that would dwarf *Spirit* and *Opportunity*. If MER was where I'd learned to lead from great mentors like Dara, Wayne, and Richard, *Mars Science Laboratory* was going to be an opportunity to apply those lessons on a larger scale and to see what I could do with them.

Meanwhile, *Spirit,* the first of our two MER rovers, was going to be launched into space on June 10, 2003, my fortieth birthday. *Opportunity* would be launched one month later. Wayne encouraged all his people to be there for the big event, and he got us badges, even if we did have to pay our own way to Florida.

I went to the birthday launch with Ruthann and Caledonia, and we watched from the pier at Jetty Park in Cape Canaveral as the rocket lifted off. I found it so compelling that, years later, when we launched *Curiosity,* I arranged funding to bring a woman from our team back from a work trip to Europe so she could be there to celebrate with the rest of us.

Seven months after *Spirit* launched, the two spacecraft were poised for their date with destiny, and we were going to get to see if our hard work had paid off.

Wayne had brought several folks back onto the project for *Spirit*'s landing. We had two spacecraft flying, so we wanted to be ready to learn whatever lessons might be gleaned from the first landing (or landing attempt!) to make the likelihood of success for the second landing, in a few weeks, as great as possible.

On January 3, 2004, the EDL team were sitting in a row in front of our laptops in a huge area we called the war room. Screens filled with data surrounded us. As we went through the EDL sequence of events, a cheer would erupt with each successful benchmark. Parachute deploy—Yay!!!! Heat shield separation—Yay!!!!! And so on. It felt surreal. I was tense but also emotionally detached. After all the time spent in anticipation, how could the big moment be happening at last?

When *Spirit* landed, the room exploded into chaos. The raucous excitement was amped up a bit by the fact that some folks had snuck in whiskey and champagne. Wayne Lee, like the good coach that he is, gave a great speech. "No one can ever take this away from you," he said. "What you've done is incredible. . . . We've just changed the history books, and you should be very, very proud."

It was a great night. We had just landed the biggest rover ever put on another planet.

But in the war room, the EDL team was mostly hidden away from the publicity and the public pronouncements. Our guys dribbled down one by one to the von Kármán Visitor Center for the press conference, and one by one they were turned away because of overcrowding. This event was for press only. So we simply left, with no press conference catharsis.

A few hours later, *Spirit* would start sending back its first color pictures of the Martian landscape. As I looked at those images of the undulating terrain of red sand and black basalt rocks, I had a slight sense of anticlimax. We'd seen this landscape before. This was the Mars we'd seen from *Viking* thirty years earlier.

Even so, I still had too much work left to do to indulge in a psychological letdown. During the many months that our two rockets had been traveling toward Mars, I'd been totally immersed in building the rover for *Mars Science Laboratory,* but Wayne invited me to rejoin the MER team for landing operations—an odd term, given that during EDL the spacecraft fly themselves, but we humans need to be around to look at the data coming down and interpret it. More to the point, this was a project with twin rovers coming in three weeks apart. Because we had, in theory, two identical landing systems, we set up a plan to do rapid analysis of the data from *Spirit's* landing to try to understand whether it had all gone the way we'd thought it would or if there had been anomalies that might lead us to make some adjustments for *Opportunity.*

One issue was how frequently we would use Miguel's DIMES and TIRS systems for steering our spacecraft rockets and under what conditions.

Wayne had invited the whole team to weigh in during "tuning" sessions, during which we'd analyze simulations to see which steering actions cured which ailments. There might be twelve people in the room and four or five options, and everyone voted, but it wasn't a democracy. Wayne would make the call, and his clear decision making was a great gift to the team. There always needs to be a general, "the decider," and Wayne taught me how to do that, to close the debate. This allows team members to love or hate the boss but to spare one another the blame for unpopular decisions.

Because of bandwidth limitations, the data from *Spirit*'s landing hadn't all come down at the same time. For a while we got every fourth data point, and then every other data point, and finally full data, though in multiple chunks. Merely assembling all this information into something clear and useful was a slow-rolling and cumbersome process.

Bob Mitcheltree was a very gifted engineer and close personal friend from NASA Langeley who had left civil service to come work at JPL. He and I were in charge of looking at the parachute-descent portion of *Spirit*'s EDL. This included the performance of the lander and bridle and the descent-rate limiter, as well as the winch and brake mechanism used to regulate the lander as it drops down to the end of its tether. We had anticipated that it would take between nine and eleven seconds for the lander to descend. When we looked at the first data set, which leaves out every other data point, we saw a tiny little blip, a little wiggle in the graph, at eleven seconds. We'd expected a blip when the tension in the bridle arrested the descent— we call it snatch force—but this blip seemed too small. It looked almost like a little noise squiggle. But it was there, right when we thought it would appear, so we moved on, though still thinking that it was a little mysterious that the snatch force was so low.

A couple of days went by, and then we got the full data set—the complete data with no gaps—and we were in total shock. There before our eyes was a clear signature of snatch force at seven seconds. It was the strongest we'd ever seen, consistent with the increased speed that the lander would have been traveling to get to the bottom of the bridle that fast. We kind of freaked. Well, freaked quietly and stoically, the way steely-eyed space explorers freak. This was potentially enough force to damage hardware. It might even be enough snatch force to break the bridle. It hadn't snapped this time, but we had another landing of the same system in three weeks. It was time for another tiger team.

I led the investigation, and given the urgency, I had access to just about anybody at JPL or NASA. We had a bird flying, and we had about a week to figure out what the hell had happened the first time. In plainest terms, was there a fatal flaw that might kill the second landing?

We focused on the descent-rate limiter and its centrifugal brake, which in the past had relied on a complex organic brake pad that used cashew nut shell oil as a lubricant. In the run-up to *Pathfinder,* the EDL team had put a few of these brakes in a vacuum chamber, pumped them down to Mars pressures, kept them like that for a few months, and determined that long-term space exposure—meaning vacuum and extreme cold—had no effect. We'd never done similar studies on the descent-rate limiter for MER. We thought there was no need, because we'd been assured by the vendor who'd sold us the brake pads that they were the same as *Pathfinder*'s, and we were leveraging the *Pathfinder* data, so that ruled out space exposure as a problem. But then, despite all the vendor's assurances—even the part number was the same—we found that the cashew nut shell oil was no longer being used in the formulation. So, despite assurances, these were *not the same* brake pads that had been used on *Pathfinder*. In other words, all bets were off.

I had dozens of people on the problem, and we took a mock-up of the MER lander to the 25-foot-tall JPL vacuum chamber up the hill. We kept it going 24/7, pulling out all the air, cooling the chamber down to Martian temperatures, then letting the lander descend again and again. This allowed us to replicate the lander's weight dropping down toward Mars under Martian conditions, but it wasn't cheap. We spent $650,000 on liquid nitrogen for the cooling alone.

The total expenditure for this matrix of tests was $1 million, and it seemed that my hypothesis about the cashew nut oil was panning out. But then a more senior mechanical engineer, a friend and mentor of mine, Jim Baughman, said, "Why don't you just do a double-blind test to double-check," meaning: Do the same brake test at Earth ambient conditions, and see what happens.

We started this test at 10 a.m. and finished at about 4 p.m., and I had to brief Pete Theisinger at 6 p.m. I'd written up a bunch of slides documenting our cashew nut shell oil theory. I'd actually pre-briefed several people, and everyone was on board with my compelling and cogent-sounding argument. But then, about forty minutes before I was supposed to present to the entire team, including the project manager, as well as Gentry Lee and lots of other board members, the double-blind check came back, and it confounded the data. The results were the same under Earth conditions, which completely undercut the space effects/vacuum/absence of cashew nut shell oil hypothesis. It was an exercise in humility as I went through my slides in the big conference room with all the TV screens and projectors, only to end with confounding data.

When Pete asked, "So, Adam, what do I do with all this? What conclusions do I draw?" I was totally spent. I had hit a wall. My model was wrong, and I did not have a plan B. I decided to come clean.

"Pete," I said, "I've got nothing. I don't really know what to tell you to do. But I think we should try to tune EDL to get more time on the chute, because whatever is going on, more time will help."

By that I meant that, for the *Opportunity* landing, we should deploy the parachute a little sooner, which would give us more deceleration, which would mean that the rover's descent down the tether would happen at lower G levels and thus create less snatch force. But opening the parachute sooner meant opening while going at a slightly higher speed, which, after all our angst about the parachute, increased the stresses.

Pete asked Wayne and Rob Manning, the flight system engineer, to look over my recommendation, and they agreed with me. This led to a frenetic three-day effort to see how far up the time line we could move the parachute deployment.

Ed Weiler, the NASA associate administrator for the Science Mission Directorate, was at JPL for the launch, and I wound up having to brief him about why this was a good idea. Making a change this late in the game was a big deal, and the concern went all the way up the chain of command.

Once again I was peering into the chasm where the bedrock truth of the hard data—the math and the physics—did not hold the answer. They are incredibly precise, but they are merely the tools we use to build our models, so you can't blame them when the model fails. In this case the only way forward was to exercise those particularly human capacities of engineering judgment, and maybe Spidey sense, as well as an activity called risk leveling, which is the essence of Entry, Descent, and Landing.

EDL is probably the riskiest part of a space mission, so managing and reducing risk is the name of the game. As the engineering development progresses, risk leveling means that we look for concentrations of risk and then we look to reduce the challenges and risks in that area. Usually this takes place mostly in development, but when you're flying a bird and you find an issue, you have to do what you can, exploiting your evolving knowledge of all the high-risk items that are coupled together and trying to balance those across all the connections.

In the end we raised the parachute-deployment dynamic pressure (the apparent wind pressure at which we open the chute) from 830 to 835 pascals, and that gave us maybe an additional ten seconds of deceleration on the chute, which lessened the load on the bridle.

By the time January 24 rolled around, I felt as if I had lived two or three lives since *Spirit*'s landing just three weeks earlier. As we settled in at our consoles in the EDL war room, primed and ready to follow the streaming data, we all remained on the edge of our seats, hearts in our throats. We had stuck the landing with *Spirit,* but the intervening weeks had reduced our confidence, and the odds with Mars landings were still worse than fifty-fifty. The confidence that we were going to simply roll through another successful landing just wasn't available.

Again we cheered as we passed each benchmark. Again we held our breath waiting for a signal from the spacecraft telling us that the suspense was over. Or maybe no signal, telling us that there had been a catastrophic failure.

When we heard the signal, we lost our shit. Everything had worked out perfectly! Bourbon shots for all! Two victories in three weeks! This called for a long succession of toasts, and maybe a good night's sleep. There were no more landings to prepare for. We were done!

When we saw the pictures that came down from this second landing site, Meridiani Planum, we were over the moon—or maybe Mars. In these images the landscape of blue surface with white outcroppings was so totally different. This wasn't the same old Mars from *Viking*. This was new, this was discovery, this was exploration!

I took a moment to savor not only these new Martian vistas but the great feeling that had developed with this team over the years. I thought about knocking back the bourbon in that Holiday Inn in Idaho when we were testing the parachutes, and those early-morning runs along the mountain trails to clear our heads. I thought about traveling to Ohio in that blizzard with Dara. We owed a lot of EDL's camaraderie to Wayne, who had been just a great mentor in terms of team building. Through his great talents as a listener and a leader, Wayne had drawn everyone up to his level and made it a very nonhierarchical unit.

The night of the *Opportunity* landing, as one toast followed the next, some of our folks had the idea to head down to von Kármán to try to get into the "official" celebration, the press event for media and dignitaries. I thought we needed a better strategy for full enjoyment of our victory this time around. No more letting my shy and unassuming colleagues be turned away from the press conference. So as we had a few more rounds in the war room, I grabbed the microphone and announced that no one was to leave. "We are going to take a few more toasts, and then we are going to head to the press conference *together*."

As we gathered outside our operations building, I was a bit taken aback at the size of the team that had been fully assembled and fired up. We began to make our way toward the auditorium, and as we did, the guys began to chant, "E D L, E D L . . ."

A pair of armed JPL guards stood at the entrance to the press room, and when they heard us and saw us, they clasped their hands over the doorway and said, "Can't let you in. Fire code—we're at capacity."

I turned and looked back at all my guys, and I saw the pride and excitement and anticipation in their faces. There was Miguel, really letting go, and Wayne Lee, and Tom Rivellini, and the teams from Langley and Ames. We had just done the impossible, successfully landing twice in three weeks on a planet that had killed two-thirds of the landing attempts to date. How could they not let us in? I turned back to the guards and said, "I'm sorry." Then I gently grasped their wrists, pulled their hands out of the way, and opened the door, and the whole team stormed in right after me.

There was a stage and a dais set up for the cameras, and I just wanted to get everyone into the room, but as more of us crowded in, the only space for us to migrate to was down the aisle, moving toward the stage and the dais. So I took us in that direction, everyone still chanting, "E D L!" over the chatter of the camera shutters.

We were now in full view, and everyone not in our group was taking a step back. We were poised as if we were going to march up onto the stage, but that had never been the plan. In fact, there was no plan, but the stage now seemed like the only place there would be room for us. So I continued down the path as it naturally led down the aisle and across the front of the stage, where we high-fived all the officials, starting with Ed Weiler, who leaned down to slap my hand. "Please excuse our exuberance," I said. So it became this spontaneous, organic conga line of high-spirited cele-bration, which ended with me standing awkwardly next to Al Gore, who was standing awkwardly in the back.

This "storming" of von Kármán became a small legend in JPL circles. Years later guys would come up to me and, choking up, thank me for doing it, even though it had been entirely an accident.

After the dust had settled, I was still in the back of the room, and I noticed Dara.

He saw me and said, "You did good."

I said, "I don't know what just happened. But I know I have to thank you. This has been an absolutely incredible fucking ride. You gave me a great opportunity, Dara, and I really appreciate it. I look forward to learning more from you."

He said, "Adam, my work with you is done. You're ready."

But his confidence was premature. I still had a long way to go.

Chapter 8

THE LEAST UNACCEPTABLE SOLUTION

AS I WAS WRAPPING UP MY INITIAL WORK ON MER, MY STOCK was pretty high at JPL. I was asked to consult for the *Mars Science Laboratory* (MSL) landing team; my capable colleague Jeff Umland already had the EDL job. And anyway, I wanted to do something totally different: I wanted to be in charge of the surface mission.

MSL was going to employ a new class of rover, the behemoth that would come to be known as *Curiosity*. Being in charge of design would mean integrating the work of all the subsystems engineers, and it would be a big step up administratively, comparable to Wayne Lee's role in EDL for MER.

So I called Dave Woerner, the flight system manager for MSL, and asked him, essentially, "What do you think? You wanna offer me a job?"

In keeping with JPL's custom of letting people "audition" for bigger roles, Dave had me come in to talk.

I didn't realize it until after the fact, but this was my first "executive interview," and it caught me flat-footed. The role I was applying for was at the level of project leadership. Up until now my career had more or less ridden the thermals of events, and I'd never been officially recruited for anything. Now Dave was asking questions that required an executive awareness: "The relationship between flight system management and project management on MER. How did that go? What were the strengths and weaknesses?"

As far as I was concerned, project management was Pete Theisinger and flight system management was Richard Cook. I'd never really thought about these things in generalizable terms. So my answers were not exactly compelling and insightful. We chatted for a while, but he didn't know who I was, and I didn't know how to pretend to be what he wanted.

When we were done, he smiled and shook my hand and politely said something that felt like "Don't call us, we'll call you."

Shortly afterward I ran into Richard. He never expressly told me this, but he gave me the impression that Dave had phoned him to ask about me, and that Dave was going to pass until Richard turned him around.

And I can't say that I would have blamed Dave if he had said no. This was my attempting to self-authorize again, big time. Given that I knew next to nothing about rovers and little about general spaceflight systems beyond the mechanics of EDL, the job I was asking for was a stretch, but apparently Richard thought I was up to it.

For Dave, who'd never worked with me, there was nothing obvious that would make him see that potential. But thanks to Richard's endorsement, I got the job.

The concept of the *Mars Science Laboratory* mission evolved from previous plans that had been frustrated. In 1999, when we lost *Mars Polar Lander,* two big ideas were on the table, but the lab was still reeling. One plan was MER; the other was the Mars Sample Return mission, with a huge rover, to be launched in 2003. According to the original scenario, the same lander that would carry the rover to the surface would also carry a launch vehicle to take soil samples back up to Mars orbit, after which a second vehicle would fly up and retrieve the launch vehicle.

With the decline in confidence that followed the loss of *Mars Polar Lander,* though, the latter mission was put on hold. Instead the lab began planning a less ambitious mission, focused purely on the technological issues of landing a big rover. This mission was called Mars Smart Lander. But then it became clear there was no way we were going to be allowed to spend close to a billion dollars to do something so grand with technology development as the only payoff.

Then, in 2003, Mars Smart Lander was rebranded as *Mars Science Laboratory.* The rechristened mission was still a way of developing technology, but now it was also going to be packed with science to deliver more bang for the buck. It was still going to have a rover that was huge in comparison with anything that had gone up before, but that rover would carry an analytical laboratory to decompose the geochemistry of Mars rocks. That on-site analysis offered the possibility of a major advancement in our understanding of the early Martian environment and whether it would have been habitable for life. That was the rover design challenge I had my eyes on.

An architecture was already in place when I came on. We were going to use the six-wheel "rocker-bogie" mobility system (a wheel arrangement that maintains the load-sharing features of a three-legged milk stool, keeping all six wheels on the ground) that we had just used successfully on *Sojourner, Spirit,* and *Opportunity.* But there was also a lot of new in it, which meant that once again we were on the steep end of the learning curve.

I absorbed everything I could about the computers, fault protection, radio systems, surface navigation, and mast-mounted cameras. I remember early meetings where I'd be listening intently to conversations and trying to infer the meaning of a whole new set of acronyms. It felt like a kind of jam session where I didn't know the tune but I was playing along, faking it. Each time we repeated the verse or chorus, I got better and more familiar with the melody.

But as I made the transition from being a mechanical engineer working on EDL to being the guy in charge of all the engineering of a rover system, I made a mistake that cost us dearly.

Operating without Dara's intuitive genius, the rover team had raced ahead with a solution for the problem of obtaining and processing geological samples for scientific analysis. It's called SA/SPaH, for Sample Acquisition, Sample Processing, and Handling. The operating assumption had been that we would need to drill core samples, spit the core out onto an inspection tray, then survey the materials on the tray with onboard instruments to decide which were worth processing for further investigation. After that cursory inspection, we would crush and powder the core and distribute the powdered product to the analytical instruments on board.

This was a huge new development and one I wasn't able to critically evaluate. I merely accepted the preexisting plans. I didn't dig deep enough for the truth, and I didn't hold on to the doubt. But I didn't leap ahead with solid engineering judgment, either. Amid a sea of new ideas and acronyms, I was scared, and I felt the need to prove that I had what it took. I was too eager to have an answer, too eager to move forward so that it would look as if I knew what I was doing and Richard hadn't made a mistake in recommending me for the job.

After much revving of our engines without getting anywhere,

we realized we didn't need to take core samples at all; we could get by with a simple hammer drill, which would create powder that would be dragged up the flutes of the drill bit and prepared for analysis. If we had simply held back a little longer and thought through the process more deeply, we could've saved ourselves an enormous amount of trouble. That path of sticking doggedly to a solution without continuing to ask if it was the *right* solution cost us at least two years on the sampling system alone. We burned out several good engineers, and we set ourselves up for a terrible slog throughout the remainder of the mission. And it was nobody's fault but mine. I was too afraid to appear indecisive, too concerned with making measurable progress, and not willing to question.

I feel that many leaders walk a difficult line between these two poles. On the one hand, we have to question every assumption, dig into every truth. But questioning can also go too far. So when does holding on to the doubt cross over into paralysis by analysis? And when does an attempt at being intuitive and instinctive lead instead to just being half-baked and wrong?

Breaking the matter down, understanding your state of understanding, and keeping your mind free and purely focused on the matter at hand—not yourself—is the only way to strike the right balance between consideration and action. You have to understand the problem in terms of those Rumsfeldian logic boxes: the known knowns, the known unknowns, and the unknown unknowns. You have to get your arms around everything you understand about the task at hand, along with everything you know you don't know. But you also need some sense of the magnitude of what you can't even directly sense. You have to be personally invested, but you also have to be detached. You cannot be bound up in concerns about your own performance or what people might think of you. You cannot be waiting for some set of information that, like Godot, may never come, and you cannot be wedded to a position that will not change when the information does arrive.

I know it all sounds very "one hand clapping" mystical, but it's the only way you can bring your full consciousness to the problem.

While *Sojourner* was the size of a bread box and *Spirit* and *Opportunity* the size of big lawn mowers, this new rover, *Curiosity,* was the size of a MINI Cooper and would carry ten times the mass in scientific instruments of any previous rover.

When I came to the mission, this big hunk of rover had two radioisotope thermoelectric generators (RTGs) for power and two robotic arms. It had identical computers—one main and one for backup—with memory armored to withstand the radiation of space, as well as self-monitoring systems to keep tabs on factors like temperature. My first task was to make it smaller so that Jeff Umland could land it.

The team dug in, and we brought the weight down by almost half—from 900 kilograms to 500 (about 2,000 pounds to 1,100)—in the process getting rid of one of the arms and one of the RTGs. The inevitable laws of spacecraft development meant that, over time, the rover would regain the weight, but that's why we wanted to start small. The rover we launched into space in 2011 was very close to the gram weight of the rover I inherited when I joined the project in 2003.

But even while we worked to bring the weight down, I continued to argue for big dimensions—a big wheelbase and big wheels. My reasoning was that you don't want to get stuck in the middle. A small rover fits between rocks. A big rover rolls over rocks. A medium-size rover loses on both counts.

The six-wheel rocker-bogie system kept the wheels on the ground in extremely rough terrain. Each wheel had its own motor, and the two front and two rear wheels also had individual steering. The cleated wheels were twenty inches in diameter and independently actuated and geared for rolling through soft sand and scrambling over rocks. Each of the four corner wheels could be steered independently, either for a wide arc or for turning in place. The center of gravity was kept low to withstand a forty-five-degree tilt without flipping over, but it also had cameras and onboard software to protect it from driving into anything dangerous. We had mast-mounted cameras to take panoramic images, plan our drives, and even observe our tracks on the surface to measure movement, as well as body-mounted hazard-avoidance cameras (hazcams) to provide close-ups of any impending obstacles. To these engineering cameras we anticipated adding science cameras on the mast to further enhance the rover's imaging capability.

But as we designed our rover for optimal exploration and data gathering, the bigger problem still looming over MSL was how to land this hefty vehicle safely on Mars. We wanted access to the widest range of scientifically interesting landing sites—not just the flat, even surfaces that could accommodate a legged lander—and as we navigated toward the landing site, we wanted the maximum precision and control to minimize the risk of some unknown hazard as we made contact with the surface.

Natural instinct pointed us toward adapting solutions that had proven themselves with recent success, which meant some combination of parachute, retro-rockets, and air bags. But we knew that, unlike with the rovers for *Pathfinder* and MER, we were not going to encase a rover this size in a tetrahedron-shaped lander for its final descent. It was just too big. So our situation was not exactly "back to the drawing board," but it was "back to the shelf" where we stored all the plausible ideas that might be reworked to suit present needs.

Despite all the air bag issues that had given us headaches on MER, the MSL EDL team at first tried to make even bigger air bags work to cushion *Curiosity*. Tom Rivellini invented a way to put the bags onto the rover directly, with no intermediate structure, and a new way to get the rover out of the air bags once we'd landed. Unfortunately, *Curiosity* was just too massive, and we could never get the fabric of the bags strong enough. With that limitation, we could never make the whole system close.

As you add more fibers to the bags to gain strength, you get thickness, but the material is no longer as efficient at evening the load across all the fibers. Eventually the fabric stops getting stronger—it just gets heavier. In the end *Curiosity* was too big for any fibers known to humankind, which meant that air bags just were not going to work. Unless, of course, we could find other means to substantially slow down the lander's descent before air bag deployment. So the EDL guys kept looking for those other means, and I continued to consult whenever asked.

The people who manage the early development of a project are often chosen for their ability to keep things moving without being absolutely certain they're moving in the right direction. Early on it's enough to simply be advancing, exposing ideas, stirring up possibilities. Later you need people who are going to be brutal on sorting through the ideas and incredibly precise on directionality.

The initial team for MSL kept things moving, but many of them had never been tested in a hands-on way. In some cases inexperience had made them timid, while in other cases it had made them overly aggressive. There were ideas being developed that were too out there, as well as ideas that weren't out there enough. And on EDL especially, leadership was unusually distributed across too many of the different NASA centers that had been brought in to collaborate. The fact is, some of the staff at these centers had never been through mission development before, and some of the centers had not done a real mission in twenty to thirty years.

One of the overly ambitious ideas was a complicated radar for guiding our oversize rover to the surface. It was an electronically steerable system designed to look for hazards, then focus in and get more information. This is called active phased array radar, and it's what the Navy uses with those wide, flat panels atop the radar towers for Aegis weapons systems. The system starts out with a wide beam, then synthetically focuses on objects that it encounters. But this concept was sunk by its own weight.

It was replaced by a simple fixed multibeam system that gives range and velocity but no imaging, similar to the system used in helicopters. These are available commercially, but most of the electronics technology used on Earth would fall apart in space because of radiation, the vacuum, and the extremes of heat and cold. So our options were either to take one of these commercial systems and vastly upgrade and rework it to make it "space rated" or to build one from scratch ourselves. We chose the latter option.

Another of our objectives was for MSL to fly a lifted-guided entry. That is, rather than simply flying smack into the atmosphere of Mars head-on, we wanted to fly at a canted angle that would give us lift. Lifted entry had been used for the *Viking* mission, but guided entry had been used only on Earth, with the space shuttle and the *Apollo* missions. By combining lift and guided entry, we would be able to steer the spacecraft side to side or up and down on its way through the atmosphere. This control would allow the spacecraft to land much more accurately, trimming the miss distance from +/– 100 kilometers down closer to +/– 10 (62 miles to 6). A small miss distance means you can find more spots on Mars where it's safe to land, which gives you better opportunities for science without having to spend excess amounts of time and energy driving around on the surface.

Lift is accomplished by having the center of pressure—the place where the atmosphere is pushing on the spacecraft—be different from the center of mass or gravity. You can achieve this difference either by creating off-center mass, as when you move the passengers forward on a small boat; or by creating off-center aerodynamics, which we would achieve by relying on something called a trim tab, a small flap or shelf-like extension used to create off-center aerodynamic pressure. The choice between the two approaches would have huge implications for the overall design, so the question needed to be resolved sooner rather than later.

Of course, once the entry body with its lift slows us down from 13,000 miles per hour to around 800, we still needed that other, familiar yet problematic aerodynamic system to slow us further: the parachute.

The initial plan was to use a chute with exactly the same geometry as *Viking*'s, a design that had been the gold standard for thirty years. We would leave behind the struggles with parachute design that had plagued MER and go old-school *Viking*: OGDGB (Original Gangsta Disk Gap Band). But given the added weight of *Curiosity*, there were doubts that one parachute the size of *Viking*'s was going to be able to slow us down enough. So the plan called for this replica chute to break the supersonic speeds and a second, subsonic chute to complete the job.

Like the question of a trim tab, this was an issue that required hard examination and decisive resolution before the team could really make substantial progress.

For the final kiss with the Martian surface, we also needed some kind of touchdown system. For *Viking* and the ill-fated *Mars Polar Lander,* this had been the three-legged lander. For *Mars Pathfinder* and the Mars Exploration rovers, it had been Tom Rivellini's air bags. But since *Curiosity* was so big, we needed to completely rethink how we were going to soften the impact.

A year or two earlier, the team had turned to something called the pallet, a concept developed in response to the failure of the three-legged approach with *Polar Lander.* When we did a failure analysis with an external board, we were reminded just how tippy a legged lander can be under the best of circumstances. Put an oversize, 900-kilogram (2,182-pound) rover on top of a legged lander, and it was going to get tippier still.

Instead of relying on spindly legs, the pallet lander put the rover on top of a shock-absorbing platform. The platform had legs, but mostly for stability, and they were spread out wide like outriggers. But the key idea was for the whole thing to crash into the Martian surface at about 10 to 20 miles per hour and for the platform to absorb the impact. And therein lay the problem. The propulsion system, which included toxic and sometimes explosive fuel, would be located on the underside of the shock-absorbing platform, and once we'd armored it sufficiently against the rocks that might cause it to rupture, it would be so heavy we really couldn't launch it. And that added weight was never going to make the interaction with the Martian surface any less complex. Even if we were able to get it up into space, I had grave doubts we would ever be able to confidently put it on Mars.

The team was stumped. In the summer of 2003, EDL head Jeff

Umland gathered together all available EDLers for a brainstorming session in room 201 of Building 158. There were fourteen of us, and for a couple of days we flipped through our library of possibilities, putting them up against the laws of physics and the objectives we were trying to achieve.

If we could slow the descent more with rockets, we still might be able to get air bags to work, but the air bag system would have to ride the parachutes all the way down through the rocket-assisted descent, and the parachute could make the system swing back and forth, which would make the rocket firing anywhere from less efficient to downright dangerous. That's why Mig had come up with TIRS and DIMES, those systems that quickly analyzed images to determine speed, for MER, to do a little steering of the rockets to correct for the wind-induced swaying. These were, however, by no means perfect systems.

Another idea that had been kicking around since 1999 was called "rover on a rope," which meant that the naked rover would first dangle at the end of a 20-meter bridle beneath a descent stage with rockets, then break away from the parachute entirely and continue toward the surface guided by the descent stage. But even though this approach got you off the parachute with its pendulum problem, having two bodies (the rover and the rocket-powered descent stage) connected by a tether added as much complexity as it resolved. Instead of being like a rock falling through space, you now had something more like a bolo—the Aboriginal hunting tool made of two stones at the two ends of a piece of twine or rope, which is thrown at the legs of prey to entangle them—which created dynamics of swinging so scary that folks concluded (incorrectly) that the system was unstable. So "rover on a rope" was set aside.

Even though we were still tortured by the possibility of swaying back and forth like the pendulum on a grandfather clock, we kept coming back to the idea. Swaying while firing propulsion rockets meant we couldn't fly with the agility and precision we needed, but we liked the idea of getting the parachute, which was like a big, unwieldy sea anchor, out of the picture. Our search began to focus on ways to reduce the swaying of two bodies tethered together and falling to the surface while we slowed their descent with the firing of rockets. The question led to a great session of ideas bouncing around the room and many conversations and debates going on in parallel.

Tom Rivellini, mechanical systems engineer for MSL EDL, stood at one of the whiteboards as everyone in the room considered the problem. The ideal would be to have a short pendulum while we were flying under the rockets and a long pendulum as we approached the ground. But we were stuck in the paradigm of MER and *Mars Pathfinder*, where the rover was lowered on a bridle while still on the parachute and that bridle was of a certain fixed length.

We all knew that whether you're talking a grandfather clock or spacecraft, the frequency of a swinging pendulum is the square root of gravity divided by the length of the pendulum. The thought seemed to be percolating throughout the room: If you take that formula and shorten the length of the pendulum to zero, the frequency goes to infinity—meaning that it no longer exists. No pendulum, no swinging.

I remember Jeff Umland and I looking at each other and saying, "Drive the frequency to infinity." We looked up, and Tom was drawing exactly that concept on the whiteboard: The two bodies—the back shell and rover—flying together off the parachute but not yet separated by the bridle. It seemed as if the whole room had suddenly had a single thought in perfect unison. You kept the descent stage and the rover together until you'd done all your agile flying and killed all your horizontal motion. Then, and only then, when you were in a perfectly vertical descent, assisted by the rockets and maybe 5 meters (16 feet) above the surface, you lowered the rover on its rope. If we used our rockets properly, this would allow us to touch down so slowly that there'd be no need for air bags. The descent stage with its rockets would then disconnect and fly away . . . and voilà!

Thus Sky Crane was born, or at least proto–Sky Crane, which was very different from the system we'd ultimately send to Mars. Even so, the important intellectual leap had been made: Don't deploy the rover below you while you're still on the parachute, or even during most of rocket-assisted descent. Wait until the very end of flight. In a single collective thought, we had all, in some sense, graduated and left the intellectual territory of *Pathfinder* and MER.

For every one thing that could go wrong with air bags, proto–Sky Crane had a thousand, but somehow we felt these problems were manageable. The basic physics was on our side. Unlike the air bags or the pallet, which were at the mercy of the particular terrain we happened to be landing on, the Sky Crane lowering a naked rover could use the rover's own wheels and suspension, designed to conform to a rough and uncertain Martian surface, as the touchdown system.

We emerged from our brainstorming session feeling that, for all the problems of this crazy idea (and, truth be told, we thought of only one one-hundredth of the problems on that first day), it was better than the pallet and the air bags we knew wouldn't work. This approach had the flying attributes of a single body and the landing attributes of a double body separated by the bridle. This would give us maximum agility as we came off the parachute and navigated toward our landing site, and maximum stability as we neared the surface.

There is an old engineering saying: "You can't push on a rope." This meant that the descent stage wouldn't be thrown off by the interaction with the surface. All the landing event could really do was shrink the magnitude of the tension in the bridle, nothing else. In addition, our bridle would keep the rockets away from the surface to prevent pressure blowback and soil disturbances that could create instability and have us landing in a self-made crater. Even though it looked crazy, we thought we could make it work.

Of course, the trouble with crazy ideas, even when they're workable, is that to most people they still look crazy. Fortunately Mike Sander, the project manager at the time, gave it a nod without too much argument. But much tougher critics waited for us up the road.

I went back to working on my real job, which was the rover, but for the next year and a half, I continued to consult with EDL from time to time, as did Miguel San Martin—one of the toughest critics I know.

Proto–Sky Crane started out with a single bridle, which caught Miguel's attention almost immediately. A single bridle would require perfect balance, like a model airplane hanging from a kid's ceiling by one thread—in this case, a thread made of Kevlar.

Proto–Sky Crane also envisioned the rover descending on a mechanical spool inside the rover itself. The descent stage, which was seen as being at a constant thrust, would react naturally to the release of weight as the rover hit the surface, and the line would simply come off the end of the spool. Then the rockets would carry the descent stage off a safe distance, where it would crash.

To succeed, that single bridle connecting the descent stage and the rover would have to go right through the center of gravity of both, or the rover would swing and tilt as it was lowered. And with the descent-stage rockets at constant thrust, there would always be hiccups to make that swinging and tilting worse.

Over the next year, the EDL team moved from a single bridle, with the spool on the rover and the descent stage hovering at steady thrust, to a triple bridle, with the spool located on the descent stage and the descent stage continuously moving toward the ground. The rover and descent stage would be 7.5 meters (24.5 feet) apart and moving at a continuous rate of ¾ of a meter (2.5 feet) per second straight down toward the surface. We played with different touch-down logics, but all our ideas had the virtue of not requiring that the Sky Crane sense the touchdown event.

Instead the descent stage was commanded to continue down at a constant speed, with an internal logic similar to a superego watching the throttle settings. Meanwhile, clock springs would reel in any slack from the bridles. As soon as the rover's weight was off the bridles and deposited on Mars, the stay-at-a-constant-speed throttle settings would decrease by about half, which the superego would detect. After a second or so (an eternity in spacecraft computer time), the computer would confirm that the rover was on the surface, and then pyrotechnics would cut the cables and the descent stage would fly away.

This was 2005, and only 10 to 20 percent of the funding for *Mars Science Laboratory* had been committed. There were still challenges ahead in getting NASA fully on board. With a commitment review coming up, it made sense to bring on the team that had just landed two rovers on Mars—which meant the guys from MER.

Pete Theisinger, project manager of MER, and Richard Cook, MER flight system manager, came over in the same roles for MSL.

Pete would beat on the project for a while, then step up into the JPL executive council, leaving Richard as project manager. Richard would then bring in Matt Wallace, the MER ATLO manager, as flight system manager.

Pete was not a man inclined to passively accept far-out ideas, going along just to get along. As soon as he heard about the Sky Crane concept, he convened an external review board.

Pete hired the Aerospace Corporation to bring in noted experts, and in September 2004 more than forty people met for two days at the DoubleTree hotel in Monrovia, California, to go over the plan frame by frame. This review board included Harrison Schmitt, the former senator and *Apollo 17* astronaut and the last man to set foot on the moon. We had helicopter test pilots from Sikorsky Aircraft with experience in long-line logging operations; Robert Ingoldby, the Guidance, Navigation, and Control lead for *Viking;* Robert Dolan, the GNC lead for the Boeing Clipper; Hal Doiron, *Apollo* lunar module engineer; and on and on.

When the review board delivered its report in January 2005, it gave the concept a clean bill of health.

For the time being, Sky Crane seemed like an acceptable design choice or, as Gentry Lee often said, "the least unacceptable solution." It did, however, add to the very long list of miracle-like inventions that had to go right to make this landing happen.

Chapter 9
PUZZLE PIECES

NOT LONG AFTER THE TWO-DAY CONFERENCE AT THE DOUBLE-Tree, Richard Cook and I attended a meeting run by Matt Wallace, the flight system manager. The purpose of this weekly gathering was to bring together all the elements of *Mars Science Laboratory* and talk about progress, how the design was converging, and how to address issues yet unresolved.

Meetings like these are standard practice with projects at JPL. The life cycle of a mission begins with a concept and a proposal, then a mission concept review (MCR), followed by a preliminary design review (PDR) and a critical design review (CDR). At this point we were past MCR and a year away from our PDR.

The flight system meeting was at JPL, and Matt sat at the head of the table in the conference room in T-1728, the trailer that was project headquarters. Richard, as project manager Matt's boss, didn't want to intrude on Matt's authority and sat unconspicuously among the wallflowers in the row of chairs around the perimeter.

I knew something was up when Matt kept asking questions about EDL and soliciting my input. The primary issues on the agenda were the parachute design—one or two?—and how to achieve lift—weight distribution or the trim tab? Matt raised some point, then turned to me and said, "So what do you think?"

I began with "Well, it's not my call . . . ," then went on to offer my unvarnished opinion. It felt a little weird to be drawn in this way, and I was concerned that they were undermining Jeff Umland, the guy actually in charge of EDL. But then it occurred to me that both Matt and Richard were too smooth to do that. I had to assume there had been earlier conversations with Jeff that I'd not been a part of. That's the only way I could account for their being so free in broadcasting their confidence in me.

The whole discussion went this way. As each topic was brought up, I would be called on as if to deliver a final opinion.

The meeting ended, and people were gathering their notes and folding their laptops.

Matt said, "Adam, can you wait? Jeff, you, too."

Everyone else filed out except for the three of us, and Richard.

"So we'd like you to take on EDL," Matt said.

Even with all the attention paid to me in the meeting, I was still slightly stunned.

"What about Jeff?"

"We've got other plans for Jeff, other things we want him to do."

This explained Jeff's calm demeanor as he stood by and listened to our exchange. In fact, they wanted him to be chief mechanical engineer for the whole project.

"What about the rover?" I said.

"We're not too concerned about the rover. We've developed rovers in the past."

"Yeah, but they were less complicated."

Richard gave me a nod. "Thanks. But we can handle it."

We went through a few more exchanges, and by the time I walked out the door, I had a new job. This was the antithesis of the way I'd been brought on to EDL for MER. That first time I'd been sort of slipped in the back door; and even when I came to MSL to work on the rover, there had not been such a definitive start. Here the intent to change horses was crystal clear and right out front.

I went back to my office, took a deep breath, and said, "Giddyup. Game on!" I was taking on end-to-end design and technical development of the landing system. This meant coordinating all EDL elements with the efforts of the delivery organizations, as well as evaluating and managing risk assessments and communicating them to the larger project and NASA management teams.

And certainly with Sky Crane, there was also some selling that remained to be done. They were bringing me in not only as the reckless pursuer of truth but as front man for an idea that was not necessarily going to be applauded by everyone on first hearing.

On both counts I felt comfortable and in a good place. Unlike the rover job, which had been a big stretch, this was more in my wheelhouse. I knew the players and understood the fundamental challenges we would face—or at least I thought I did.

Mars, Dead or Alive, the *Nova* documentary about MER, had aired recently on public television, and I'd scored enough screen time that people I'd worked with for fifteen years started to look at me differently, letting a filmmaker's ninety minutes color their own opinion. Somehow the overly simplified and aggrandized image of "Adam Steltzner" as depicted on the small screen swayed them to think more of me, even though they'd had ample time to get to know my faults. It was weird, but I was not above seizing on this odd bit of human nature and using it to my advantage whenever possible.

There's a significant downside to being in the limelight, both in terms of rivalries and in terms of the resentment that comes when a few people are employed to tell a story in which thousands participated. But there is also an upside, if you can use it correctly. If a little brush with fame will make people more inclined to listen to what you say, you can drop some serious truth on them that they wouldn't necessary accept from somebody else. For the first time, I felt fully empowered to let my freak flag fly. I was going to bring my whole self to the task at hand, and the biggest part of that task was going to be leadership.

We humans are an extremely social bunch, and most of the great things we have done have been done working together in teams. How we organize when we work together can lead to greatness or disaster, and although the science of organization is studied and well developed, in my experience organizing humans together for a common goal is much more of an art—creative, fuzzy, and emotive.

I've learned to think carefully, when I'm assembling a team or welcoming a new member, about the entire cast of team members, their wants and desires, stated, unstated, and sometimes unconsciously demonstrated. I consider the strengths and weaknesses of everyone, including myself, and how each might grow or improve. With all this information, I take the traditional project organization structure for the portion of the project I am involved in and I warp it, distort it, and make it conform to the specific individuals involved. The obvious goal is to create the most productive team in order to deliver the best team product.

MSL EDL was a team made up of members from across the

nation, working at several different NASA centers, so I was faced with fitting together the pieces of a very big puzzle.

This idea of strengths and weaknesses, dislikes and desires, holds at an organizational level as well. Groups, subteams, and even institutions take on a personality, in a remarkable, complex, and sometimes unfathomable way. IBM has a different culture than Apple, which is different still from Autodesk, or Cisco or Google. Within NASA the Johnson Space Center has a different culture from Ames Research Center, which differs from Marshall Space Flight Center and, of course, from JPL. The differences among the various centers' cultures can be used to forge strong, highly functioning teams, or they can lead to disharmony and discordant results.

Finding how to help each of us as individuals, subteams, and/or institutions play to our strengths, grow toward our desires, and maximally contribute to the effort of the whole is the essence of leadership.

JPL stands outside the circle of NASA's eight field centers, in that we work for the government but are not government owned. We are a federally funded research and development center administered by a university. In that way we're more like Sandia National Laboratories in Albuquerque, which does work for the U.S. Department of Energy but is owned by a subsidiary of Lockheed Martin; or Lawrence Livermore National Laboratory, in Livermore, which does work for the Department of Defense but is administered by the University of California (among other entities). NASA has only one outlier of a research center, and that's us. We are not civil service but fire-at-will, private employees of the California Institute of Technology.

Broad generalities are always suspect, and of course there is a tremendous amount of variability and overlap, but I think it's safe to say that the security of working for the government appeals to certain types of personalities more than to others. In general the work done at the NASA field centers moves at a slightly slower pace, and the connection between the actions taken and the dollars spent on behalf of those actions is felt less directly. If you get a raise at a NASA field center, it's through an act of Congress. To get fired, it almost takes an act of Congress. At JPL all that is merit-based, in a free-market kind of way, so you really have to deliver if you want to stick around. But you can also expect a greater degree of freedom.

Since NASA's beginnings there's always been a certain amount of competition among the research centers. Each center has its own history, its own story, its own glory. NASA as a whole is strongest when each plays to its strengths. That does not mean, however, that every center does only what it does best.

For many years JPL had the complete lock on unmanned missions and the development of deep-space vehicles. In the nineties the pressures of budgetary contraction opened up more work for competitive bidding with private contractors but also with the other research centers trying to eat away at the JPL pie.

Our work was usually thought of as being near perfect, and near-perfectly "gold plated," which means that you better bring a big checkbook. But while other industrial NASA contractors may claim to be more "cost innovative," usually they just do work that's not as good for less money.

I'm sure that part of JPL's reputation for being expensive stems from the fact that, with our staff not being federal employees, our labor costs are not buried in some congressional budget line among all the other millions of federal employees. When NASA works with JPL, it has to cut a real check, for cold, hard cash, and that probably stings a bit. That's why I tell young engineers coming to the lab, "We're not paid to do things right, we're paid to do them just right enough."

That sentiment holds true for engineering at large. Take this delightful test of any engineering effort: Does your device, your robot, your app do what it is supposed to? Does your fancy space-exploring robot work as you planned? Making things too complex, too overwrought, is one of the fastest ways to fail. In our designs we look for the simplest, lowest-risk way of solving a problem. Sometimes that solution looks brutish and maybe lowbrow. But if it works, we have done what we needed to do with the least risk and usually at the lowest cost.

On MSL we were working with the Ames Research Center, near San Jose; the Johnson Space Center, in Houston; and Langley Research Center, in Hampton, Virginia. In the *Apollo* and *Viking* era, different field centers performed different parts of the task, frequently delivering different elements of the project. With *Apollo,* for instance, the command module development effort was led by Marshall Space Flight Center, in Huntsville, Alabama; the lunar lander effort was led by Langley; and the entry vehicle development was led by Ames. The vast majority of the actual work was done by industrial contractors, like Lockheed, Martin Marietta, Grumman, Rocketdyne, and so on. The net result was a fair amount of duplicated effort at great expense. The beauty of those times is that we really didn't keep track of how much we spent. We were in a race with the Soviets for the high ground, and the most important item on our list was winning. And win we did.

In the modern epoch, since 1980 or so, we have succeeded at exploring other planets and moons with a leaner approach. We succeeded in making it to Mars by having a centralized team at JPL focused on EDL success while teaming with and leveraging the expertise of some of the other field centers.

JPL had worked with Ames and Langley on both *Pathfinder* and MER, so we had a history. Both those institutions are aerodynamics experts, and they have high-quality real estate—wind tunnels—set aside for testing. Langley also had the leading experts on aerodynamic modeling, and they had helped us before with our trajectory simulations, developing the control system for atmospheric flight.

When I started working EDL on MSL, the distributed-leadership process had sneaked back into place, with a wider set of leadership distributed around NASA centers. This meant a less clear vision of what our central risks were and what our design should be. We were exploring a wide range of possible EDL design elements, and we were not focused enough on narrowing the job to the minimum-risk solution, then moving forward. We needed a central command to get down to the essence of things—the hard truth—and then to act on it. But to refocus our effort and tighten our mutual understanding of the design and where our risks lay, we first had to get beyond the clash of cultures.

Jeff had encouraged members of the JPL team to bring ideas to him, so much so that the diverse interests of the individuals involved were driving the architecture. This had been a fertile approach, but the result was that, when I arrived, there were some design ideas or technologies that were not necessarily appropriate for our EDL systems needs.

This was certainly true in what was coming from Langley, and there was strong leadership there, in the form of Mary Kae

Lockwood. She was nudging the process into line with her vision, which was aerodynamic flight within EDL.

For our previous expeditions to Mars, we had used blunt bodies that smacked into the Martian atmosphere blunt-end forward and then slowed down rapidly. For MSL our huge rover required the use of some lift to fly more efficiently through the sky above Mars.

This lift was understandably super sexy to Mary Kae and her team, because Langley had been all about aerodynamics for years, especially during the decades when we were still figuring out flying here on Earth. You know how quiet and comfortable a modern jet is? Most of that comfort and quiet comes from aerodynamic advances made by NASA in the second half of the twentieth century.

But in space it was a different story. In the decades when we had been using liftless blunt bodies to land on Mars, there had been little application of Langley's aerodynamic flight mechanics. Now our lifting and guided flight attracted all the intellectual energy that had been sitting coiled and frustrated for years.

Mary Kae was a very capable engineer and an exceptional spokesperson for these interests, and her advocacy influenced some of the design choices. The Langley team wanted a vehicle with a different shape than the ones we had flown before, and they wanted to design and build that shape. They also wanted what amounted to an aircraft control system to guide the vehicle to its target. These were both bold and nonobvious design choices, but the team at Langley was emotionally committed to them.

All engineers have lust for designs we want to see realized, visions of how cool something could be or how it would look to solve a problem a particular way. Often that "way" is something we love for the underlying beauty of the physics that forms the solution.

But then there is always that ever-practical Occam's razor, which cleaves success from failure, carving away all that is not needed and leaving "just right enough." This blade separates fact from the fiction of a dream too far.

The Langley team's vision was too influenced by their sense that, aerodynamically, we had been doing the same things with our entry vehicles over and over again, without innovation or experimentation. They were right about that, but the reason we were doing the same things over and over was because those things worked. Their dreams of innovation were not aligned with the minimum that was likely to contribute to our mutual success. So how would we find a way to help them be happy in a team where they weren't living their dream? Could we come to a mutual understanding of how they could contribute, playing to their strengths, but not letting their personal goals and interests take our eyes off the prize?

NASA doesn't do enough flight projects to forge a broad set of practical skills across the agency. Many groups at many NASA centers spend time studying what we might do if the nation were to commit the funds to do things that aren't going to get done anytime soon. For many the efforts under study that they can spend their careers on may never come to fruition. A classic example is studying human expeditions to Mars. Such an expedition is a substantial undertaking likely to cost tens of billions of dollars, but that price tag doesn't preclude the study of the different ways we might get humans to Mars, what fuels we might use, how many rockets it might take to get there, how long the astronauts would be in space, in Earth orbit, in Mars orbit, and so on. However, because these studies are almost always for mission concepts that are one or two decades away, at the very least, the studies involve a kind of abstract thinking and contemplation of design and risk that is not necessarily practical. It is only with the real prospect of having to make something happen—now—that Occam's razor comes out and cleaves fact from fiction. Only in an environment of imminent application are ideas made to compete in brutal mortal combat. Without really having to do a thing, we can't learn to accept the simple (and the potentially ugly) truths that should guide all aspects of the thing's design.

In some ways the possibility of guided, lifted aerodynamic flight had lured Langley away from the brilliant aerodynamics and trajectory simulation of blunt-body entry vehicles they had contributed to *Pathfinder* and MER. Even though there was no question that we were going to need guided lifted flight, the presence of it within our design had caused the Langley team to go a little crazy and dream too big. When they moved away from their experience base of analysis and into the realm of design, they were working in a domain they had experienced only in the abstract. Imagine a bunch of soldiers who've prepared for a deployment by playing Call of Duty on Xbox. They get to the battle zone all pumped up, and then they come up against a master sergeant who's on his third tour. "It's not a game," he has to remind them. "We're not just going in guns ablazing. First we're going to listen, and then we're going to tread lightly."

However well done, Langley's trajectory simulations didn't incorporate the kind of fears you carry in your bones if you've faced the vague and harsh realities of actually landing a robot spacecraft. What was going on with *Spirit*'s DRL (descent rate limiter)? Why didn't it work as we had expected? Why was our parachute blowing up during tests? Facing the fuzziness of unanswered critical questions like these forces a skeptical eye toward the complex, along with an attraction to simplicity.

A career of experience running simulations can make the whole EDL process seem abstract. One telling example was the Langley team members' fondness for the term *transition*—as in, a spacecraft is flying in its aeroshell, and then it *transitions* to the parachute.

Now, the "transition" from entry vehicle flight to parachute flight can be allocated 1.7 seconds in a model or a paper study. But the use of the term as a *verb* suggests that magic is about to happen. First the spacecraft is in one configuration; seconds later it's in another. It has transitioned, like a new frame in an animation. But what really occurs during transition is a series of complicated maneuvers that aren't guaranteed to succeed. During that transition we need to measure several sensors correctly, incorporate them into a navigation filter correctly, trigger the firing of a pyrotechnic mortar to deploy a carefully designed and perfectly packed parachute fast enough to get free of the spacecraft but not so fast that it melts itself in friction as it inflates, and so on.

In reality, transitions are where the vast majority of the engineering comes into play, and they are the moments in which you are most likely to die. That's why the use of *transition* as a verb is a dangerous indicator of arm's-length engagement, like that all-too-familiar phrase from political ass covering: *Mistakes were made.*

I also thought the Langley team's enthusiasm for the novel had its dangers. From the beginning of this mission, there had been a focus on EDL technologies. Even in the early days of Mars Smart Lander, we knew we were not going to be flying to the surface of Mars the way we had in the past few missions. The EDL community was excited about the chance to try new things, like guided entry, but everyone at Langley seemed *too* excited, as if they were hopped up on Red Bull. So the first actions I took were meant to temper some of the "steroidal" stuff that had been developing out of long-coiled frustration. I wanted to tone down the confidence and introduce some healthy skepticism, a measured eye, and critical evaluation.

We needed to calm down and work the fundamentals while keeping a few cans of Red Bull stashed away for when the need might arise.

One example of the fundamentals I had in mind: I wanted us to fly with less pitch, and at angles within the range where the aerodynamics were better understood and closer to what we'd flown on both manned and unmanned missions in the past. This lower pitch meant less lift and less capable guided flight. For the folks who had been waiting for decades to do something like guided flight to Mars, this was disappointing.

Lower pitch meant that we had to be willing to give up a little altitude capability and performance, an act of conservatism that I thought we should accept for multiple reasons. Safety was the first consideration. But also, with more limited performance capability, we would be less tempted to grow in mass. Later we could twist the knobs and gain back a little performance if it was truly necessary.

With an eye toward reining in some of the emotional extremes and bringing centralized decision making back to a clearer and more practical vision, one of the first things I did after taking on EDL was to set up a three-day team meeting, gathering the roughly forty people from JPL and the three centers working with us: Ames, Johnson, and Langley. We took over meeting room T-1729 in the set of double-wide trailers on the JPL campus where we would be headquartered during most of our development.

But by the time everyone had gathered at the lab, I'd come down with a horrendous flu. I had a temperature of 104 and was semi-delirious, so I missed the first two days.

When I showed up on the third day, I made a nuisance of myself, which I thought was part of my job in order to help forge this tighter, more practical vision of our EDL.

I began by asking questions, not like a prosecutor, exactly, but also not like a low-key, genial engineer. I needed the team to understand the urgency of getting to a deeper understanding. With more experience under my belt, I was able to bring more juice to it—my full passion—asking the follow-up questions that didn't let people slide by with answers that were vague and unconvincing.

"So this means x, y, and z?"

"Yeah, sort of."

"Well, help me understand. Teach me."

I was trying to be at everyone's shoulder, matching each of their steps as best I could, trying to learn what they were doing. I vacuumed up new information and chased everyone, including the aerodynamics team from Langley, up the learning curve.

One of the advantages of coming to my profession relatively late in life is that I had to run up several steep learning curves to get there. I'm still close to that experience, and I enjoyed it.

Anytime someone is brought in to help a system that's struggling, he needs to feel that the system is genuinely in distress in order to rev up the motivation to do the job. That's how he generates a sense of purpose, sensing that his value is proportional to how messed up the current system is. If there isn't a real problem, then the new guy brought in to straighten stuff out is just window dressing.

I'm sure I came in with that bias. I used to quip that the existing EDL design was like a patient with a fairly substantial disease who was shot full of adrenaline, slapped a few times in the face, given an espresso, and sent out into the world to pretend to be healthy. This was not a view of the EDL system design that was meant to make everyone on the team happy. In fact, it was in some respects designed to incite argument within the team. We needed to beat on what had been and break through encrusted assumptions and see new possibilities for things that could be better.

I needed to bring everyone into the game at full capacity. And just as on MER, full capacity demanded intellectual conflict, and for that conflict to be creative rather than combustible, we needed the team to get close.

We couldn't be like city-states, each with our own turf to worry about. We had to lead together, in a cross-pollinated way, being aggressive in our conversations while showing love and respect for the people. As for the ideas, we beat the crap out of them. Of course, this is harder to accomplish without day-to-day, physical proximity.

Some of our remote partnerships worked better than others. Guidance was being handled by a contingent at the Johnson Space Center, and it relied on the *Apollo* guidance algorithms. We used these for many good reasons, perhaps chief among them that they were bone simple. The staff at Johnson, including Claude Graves and then Gavin Mendeck, were top-notch. They knew the algorithms cold, and every time Mig and I pushed them to understand even better, they were up to the challenge, so it felt to me that the guidance job should stay with them. Ames was handling aerothermodynamics as capably as anyone at NASA could, so I let it stay there.

As to Langley, I had two concerns with their control algorithms. First, they were new to control design for spacecraft. The folks working the problem were dusting off textbooks from their college days. Second, those textbooks were about airplane control approaches, and our vehicle looks nothing like an airplane. It uses rocket thrusters rather than flaps and engines for control, and it was going to spend most of its time controlling itself in the deep vacuum of space. In aircraft the directions of motion—yaw, roll, and pitch—are highly coupled, meaning that a change in one necessarily affects the others. It also means that the control equations for an airplane, a slender vehicle whose motion through the atmosphere is dominated by aerodynamics, are complex and difficult to understand.

Our vehicle and its behavior in the Mars atmosphere demanded not an airplane control scheme but, rather, an uncoupled spacecraft control approach in which performance and risk could be calculated with greater certainty. And we needed it crafted by folks who did the job of building spacecraft controls on a regular basis.

For this reason I wanted the design pulled back to JPL, where we had been developing spacecraft controllers continuously for almost seventy years. Specifically I wanted to put it in the capable hands of Miguel San Martin and his team.

But we were also struggling with how to obtain the lift we needed to fly through the atmosphere of Mars. Our colleagues at Langley were driving the idea of trying to gain lift and control by using an aerodynamic flap attached to the outer edge of our spacecraft—the trim tab. They had tested the hell out of it, but the concept had some issues. First, the launch vehicles that get us out of Earth's atmosphere are only so big. Our rocket noses have only a 4.7-meter- (15.4-foot-) diameter fairing, and in general we build our aeroshells to make a tight fit between the largest diameter of the aeroshell and the inside of the launch vehicle fairing. Anything extruding from the surface of our aeroshell—including a trim tab— would require that the circumference of the aeroshell be diminished in order to accommodate what was being added. Which gives you less overall drag area with a higher ballistic coefficient, which makes it harder to slow down. So in that sense the trim tab was a step in the wrong direction.

But there was an even more dramatic concern, and that lay in how the irregular surface created by the tab might respond to the friction of flowing air and the heat this produces.

The aeroshell is rounded and smooth and easy to understand in a computer model, which makes the flow relatively easy to test. At the corners of the proposed trim tab, though, where the shelf would attach to the surface of the aeroshell, you'd have right angles, a distinctively different zone where the features would create large changes in the flow. Unfortunately our test facilities, the best on Earth, can only test small samples of our heat shield thermal protection material, and they cannot duplicate the complex flow patterns that the shelf feature would create. So the trim tab would give us the added risk of complex patterns of hot gas flow, and no way to fully understand the risks.

So I asked the team the questions: "What does it mean for us to have a complex, localized heating phenomenon? How is it going to affect our thermal protection material? And how are we going to test it?"

"Those are all things we have to consider," they said.

This vague response told me that they had given this only passing thought. They had fallen in love with their trim tab idea too soon. Certainly they had not gotten to the stage of beating up on it.

So I followed up by asking, "What is the aerodynamic environment in that corner? How rapidly does it change? How much can we test it? Are the facilities at Ames"—where we test our thermal protection material—"able to test at the sizes we think are important?"

"Ames folks say they can only test two or three inches in diameter," Mary Kae told me.

"And our feature?"

"Ten inches in scale."

So the length scale on this corner of the trim tab would exceed our capacity to test, which would give us no way to trust our analysis, and we would be left with a huge unknown.

By digging just a little more deeply for the truth, we'd discovered a potentially fatal flaw in the idea of the trim tab—namely, that it could allow the friction heating of entry to bore its way in and burn up our spacecraft.

Mary Kae was super smart and poised, able to answer every question, but with that exchange she could see that the game had changed. I realized that what she'd done on MSL was what I'd done back on *Champollion:* She'd self-authorized. She'd seen an opportunity to add leadership value, and she'd taken it, and my hat was off to her for that. The difference was that I had enjoyed the dumb luck of self-authorizing into a team organization structure that worked with me and my desires. Here she had led and aspired in a direction that in the end was not really going to contribute to our collective effort. The fact that her ideas weren't going to work was not readily evident to the Langley team because of their lack of real-world flight development experience. Paper analyses and scale-model tests just aren't the same as having to really build something and fly it. We needed the simplest, easiest-to-build solution, a solution that leveraged each of our team members' core strengths, and a solution that we could test to the point of absolute confidence. It wasn't fulfilling a dream; it was about making something that was just right enough.

For me, trim tab was dead after about a minute and a half of conversation, but we weren't looking to crush anything in single strokes. Part of respecting the human and looking at how we will fit the puzzle pieces together is allowing the people enough time to come to terms when an idea they hold dear has become a "dead idea walking." So we delayed making the final decision until we'd had a couple more meetings and let the new reality settle in. Ultimately, rather than the trim tab, we went with a weight-offset approach to create lift.

Through luck, historical accident, and hard work, JPL had over the years developed a great many successful missions to explore our solar system. In a very practical sense, the folks with experience at JPL knew what a good success smells like and which promising ideas aren't going to go the distance. But we'd also seen failure, and you learn a lot more from failure than from success.

To shift our team's culture, I continued to push hard with the questions, driving insight at a pace that established my leadership and encouraged the team members to question and dig deeper. Asking the right questions is not necessarily an activity associated with a struggle for control, but questioning can create an "OODA loop." The acronym, from military strategy, stands for "observe, orient, decide, act." It comes into play whenever there are recurring cycles of action and information in a competitive landscape. Whoever can react more quickly (and wisely)—"get inside the other's OODA loop"—will most likely win.

I knew that I still had plenty to learn myself, and to keep climbing the curve and accelerate the OODA loop, the EDL team cherry-picked Theisinger's concept review board to assemble our own EDL standing review. We brought on Bob Ingoldby, who had led guidance and navigation for *Viking*; Robert Dolan, GNC lead for the Boeing DC Clipper; Kevin Johnson, a propulsion expert from Lockheed Martin, in Denver; and several others, as well as the smartest engineer I have ever met, Ken Smith, from Alliance Spacesystems, in Los Alamitos, California.

Some people dread the kind of oversight our review board provided, seeing it as someone watching over their shoulders, breathing down their necks. I saw it as taking advantage of valuable, expert advice, and I saw the board members as advisers, not judges. I also find it a lot more intriguing and challenging to convince somebody who's skeptical of something I've proposed than it is to convince myself that they don't know what they're talking about. The challenge of winning them over makes the person who doesn't understand you much more interesting. So I embraced the board, tried to bring them along, and tried to listen to what they had to say in an open-minded conversation in which I had to at least entertain the possibility that I might be wrong. And I could definitely be wrong.

We brought Kevin Johnson in to help with the fact that we were redeveloping *Viking*-like, throttleable rocket engines to decelerate the descent stage as it approached the Martian surface. During one of the early reviews, he raised a question about our having the engine nozzles outside the descent stage, beside and slightly above the rover. With this configuration the exhaust would be a foot or so away from the rover's wheels, blowing a stream of super-heated ionized gas right past them. Kevin did a back-of-the-envelope calculation and said, "Seems to me you're going to have a radiative heating problem." He meant a problem like standing too close to the fireplace. It might not actively burn your leg, but it's going to be damn hot.

At the meeting we were looking at the design on a computer screen, and I said, "Thank you, we've done the math, it's okay." But I wasn't really hearing him. The rest of the team wasn't really hearing him, either. Certainly I was not giving him credit for being as right as he might be. We had really worked this problem, I thought. We were largely the team that had landed the twin MER rovers on Mars a few years prior, we had done our homework, and damn it, we knew our stuff! But mostly what was at play was the pride that cometh before a fall.

We have a process called verification and validation, whereby we double-check everything. As we went through the double-checking of the exhaust plumes, we discovered that our model underestimated the heating. We also realized that we were going to overheat not just the wheels but the entire rover. This was six months after Kevin first raised the issue.

So we had to shove the engines down and extend the nozzle. This meant that we also had to take the engines through their individual verification and validation testing all over again. In all, it took us another year to complete the changes and deal with the ripples caused by that one error. The six months we would have saved if we had listened to Kevin would have been valuable. It sucks to be wrong.

Even when you have a great team, and you are questioning deeply, and you are really cooking, you can still be wrong. You can always be wrong. I hope that I have learned that if even a whiff of possible wrongness is in the air, I should hunt it down. Try to make the most of the appearance of being wrong, until either you convince yourself and others that you really are right or you discover the error in your ways. Never leave a suspicion of error left undeveloped.

As soon as I was able, I made Mig head of EDL guidance and navigation. Miguel is a brilliant electrical engineer who understands controls, but one with a practical and simple aesthetic. Controls people often go off the deep end on optimization, which can take five times the effort to get a 10 percent greater return. Miguel has a more basic, structural aesthetic that I share, along with a dislike for optimized solutions and an appreciation of economy in design. He is always striving for just good enough. Ultimately Miguel supervised the writing of the software that lowered Sky Crane to the surface of Mars, and he was my indispensable partner in the whole effort of developing the EDL system.

By the end of 2005, we had a stable and focused team for EDL, and we were firmly committed to Sky Crane. But by this time, the idea that we were really going to do this outlandish thing had reached the top brass at NASA. That's when NASA administrator Michael Griffin sent a message to JPL that said, essentially, "Come explain this thing to me." This was ominous. A NASA administrator (the top boss) doesn't usually ask to critique a spaceflight system design. But then again, Mike Griffin was not your average NASA administrator.

Chapter 10

THE RIGHT KIND
OF CRAZY

RECENTLY APPOINTED BY THE SECOND PRESIDENT BUSH,
Mike Griffin became NASA administrator after heading up the
Space Department at the Johns Hopkins Applied Physics Laboratory. He had a Ph.D. in aerospace engineering, plus four separate
master's degrees, in aerospace science, electrical engineering, applied physics, and civil engineering—plus an MBA. His education
suggested a certain attention to detail, and that he was not the kind
of manager to skim a report and leave the rest to his underlings. He
was a hands-on kind of guy who relied on his understanding more
than on the advice and opinions of others. I think it was refreshing
for NASA to have someone who understood the nuts and bolts of the
business of space exploration. The previous administrator, by contrast, had been a career bureaucrat. Not Mike. He was respected
within the ranks.

Even so, a project being summoned to Washington to explain
itself was not standard operating procedure. In fact, I'd never heard

of it being done before. But the fact that it was Mike asking for it made perfect sense. It also meant that it was not going to be a bureaucratic discussion.

Richard Cook asked me to prepare a briefing package to present in Washington. He helped me put together my slides, along with Matt Wallace, the flight system manager, and Fuk Li, the notoriously hard-to-read head of JPL's Mars Exploration Directorate. This was my first real exposure to Fuk, and I was impressed by the quiet intensity he brought to something as "down in the trenches" as assembling slides. The exploration director manages all missions to Mars. The position was created after *Pathfinder* and is run out of JPL with oversight from NASA HQ. Fuk took a particularly conservative approach to the job. I'd never known him to smile, and certainly never to crack a joke. Looking beyond his shaggy salt-and-pepper hair, which gave him the look of a Chinese surfer dude, you noticed the measured gaze, the calm, flat poker face, as if he were intentionally playing up the stereotype of the inscrutable Asian. After I got to know him better, he'd sometimes joke with me that he cultivates this persona intentionally.

Once the four of us had the slide package put together, we brought in Rob Manning, Gentry Lee, and Dara Sabahi to beat up on it as well.

We spent about a week torturing the slides until we were satisfied they presented the right story. Then the bunch of us spent a weekend in a conference room at JPL shuffling and reshuffling them again and again. This was the most work I've ever done on a package of presentation materials, and I resigned myself to the frustration of being second-guessed and edited at every turn. Gentry and Rob played devil's advocates, listening to my argument and lobbing screwball questions at me.

The meeting at NASA headquarters was on Tuesday, and the core group of us—Fuk Li, Richard Cook, Matt Wallace, and I—were scheduled to fly to Washington on Monday. Rob, who would later become chief engineer, came along as well, because he knew Mike Griffin from contact they'd had during the development of *Pathfinder*.

On Sunday night I tossed and turned, in part because I was wondering about what to wear. At JPL we wear the label of mavericks proudly, but in DC you see a lot more dark suits. At NASA headquarters the dress isn't as formal as it is "on the Hill," but still, it's a lot more formal than it is at JPL. I wanted to show respect, but I also wanted to be myself. Eventually I settled on a tie, with a 1950s blazer, jeans with my MULE DAYS belt buckle, and cowboy boots.

I had a hard copy of the final slides with me on the plane, and I spent the entire flight working on my transitions. Here was a case in which the work truly was going to demand everything I had. Not just the technical chops to have the right answers but the presentation skills to convey them and the people skills to not lose my shit if the questioning got heated.

NASA headquarters is a couple of blocks off the Mall south of the Capitol, near the Smithsonian National Air and Space Museum. We checked into the Residence Inn across the street from NASA, then went over to the Smithsonian to wander around and see the *Viking* lander and the other JPL spacecraft preserved there. It was a poignant reminder of what was at stake. When what we did succeeded, it became part of the national heritage. When it failed, the failure was also absorbed into the national consciousness, only as self-questioning and loss of confidence. In the end, we work for the nation. Our triumphs and our stumbles are not our own but everyone's. Most of the time, when I am working at JPL, I am narrowly focused on the task at hand, on my job. That day in Washington, my task was to propose a landing architecture so unusual that it might end up in these very halls, in the nation's consciousness—provided that it worked.

But first we had to get it authorized.

Richard and I had talked about what would happen if Griffin shot us down. We could go back to a legged lander if we had to. That would restrict us to flatter sites, force us to reduce the size of the rover, and impose what I believed to be higher risks for EDL, but we could adapt and live with it if we had to.

So the fact was, getting approval for Sky Crane wasn't life and death. But we really thought Sky Crane was the best way to go. We'd been looking for the truth, and this is where the truth, as well as engineering judgment and Spidey sense, had led us. Even if it did look kind of odd.

The next morning we arrived at NASA headquarters just before nine, and by the top of the hour about thirty people had assembled in a small amphitheater with multiple video screens. Each of the eight NASA field centers would be conferenced in. JPL director Charles Elachi was at the other end of one of these video connections, with Miguel at his elbow, just off camera, ready to be my backup should I need him.

A few minutes after nine, Mike Griffin and a couple of his lieutenants walked in and took the seats reserved for them in the second-row center. Griffin wore a mock turtleneck and a blazer, and I remember thinking at the time that, in that outfit, in this room, he looked like a perfect Bond villain. The only question was whether or not the floor would roll back to reveal a shark tank meant just for me.

Griffin began in the usual way, thanking all assembled for being there. Then he turned to look up at the group fanned out behind him. "When I heard what they had planned as a landing architecture for this project," he added, "I said, 'This sounds crazy.' So I asked them to come and explain it."

His tone was friendly, but given the circumstances, I couldn't escape the feeling that there was also just a hint of aggression.

The next moment was filled with what I believe they call a pregnant pause. Then he looked at me and said, "So, Dr. Steltzner . . ."

I stepped to the podium, pulled up the first slide, and started in. The rest of the JPL contingent was in the audience just to my left, which was comforting and reassuring, but I quickly realized this was going to be like an oral argument before the Supreme Court. No matter how seamlessly you'd prepared, you could count on being interrupted at every turn.

I'd gone only about a minute before Mike broke in with the first question. In providing the background for our decision, I'd reviewed the limitations of our other options, and I'd asserted that legged landers suffer from unknown pressures as the products of a rocket plume billow up from an uneven surface.

"An autopilot can handle that," Mike said.

"Well, the bandwidth necessary is a problem," I said. "And certainly it becomes a validation and verification issue."

"No," he said, "a modern autopilot can handle it. . . ."

I disagreed. It would depend on lots of factors. What was the nature and functional dependency of the disturbance? What could the controller bandwidth handle? To my eye, it was not a simple truth that the autopilot could handle it.

I paused for several long beats. What to do? Prosecute my position, or just move on? I suddenly realized that even if I could win this debate between myself and the head of NASA, winning points wasn't the point. We were there to get approval for a single, very big idea. That's all that mattered. The case I was making wasn't something I'd cooked up overnight when I found out we were going to be called to account. This was a case that the entire team had developed over a period of years. It had been approved by our review board from the Aerospace Corporation and by our own EDL review board as well. It was all new to Mike, but the intellectual progression had been well exercised, and the weaknesses shored up. So I let the argument go—ideas should win, not people. Then I went on with my slides.

Our session went on for about ninety minutes. Mike made fewer interruptions as I went along, and judging by his expression, my best guess is that he was still not too hot on the idea—he just couldn't find the fatal flaw that would lead him to tell us to drop it, reverse course, cease and desist.

I got to my last slide, finished my summation, and then stopped.

There was another pregnant pause, after which Mike said, "Thank you very much." He made that same half turn he'd made at the beginning, then looked up into the amphitheater. He gazed around the room for a few moments. Then, finally, in what seemed to me like a tone of resignation, he said, "Well, I still think it's crazy. But it might be crazy enough to work. It might just be the right kind of crazy."

Until that instant Sky Crane had not picked up much traction with the other NASA centers. Whenever I'd asked for support—asking whether we could use the big vacuum chamber at Glenn Research Center, for instance—I'd gotten a lukewarm response, friendly but slightly evasive, dragging their feet. I think everyone thought this idea was never going to make it all the way through the review process, so why should they go out of their way for it? Now, with Mike's endorsement, stated clearly in front of every center director via video, everybody wanted to be on board. The bell had been rung by the guy on top who says we're doing this, and suddenly we were going to have access to whatever we needed.

Mike and his lieutenants stood up, and then he gave me a smile that seemed both amused and still slightly skeptical. It was if he were saying, "Okay, you made the case . . . now *you* get to go make it happen!"

The top brass exited rather quickly, others milled around, and some even came down to congratulate me.

Among these was Fuk Li, who, as I mentioned earlier, is not known for displays of emotion, either positive or negative.

He stopped one rung from the bottom of the steps so that we were on eye level. "Adam," he said, "well done."

But there was no accompanying handshake.

Then he added, "I would like to give you a hug."

And so he did.

So we got the job. Now all we had to do was *do* the job. We still had a ways to go on the EDL architecture. I knew at the time that we had some challenges in front of us, but I really had no idea. We still needed to work out how we would get our lift, through the trim tab or another method. We would continue to struggle with the parachute, as we seemed to always struggle with the parachute. But at that time I had no idea how hard the parachute would be, how challenging landing on the rover wheels would prove, and how, as we took steps to eliminate environmental risk factors on Mars, we would in some sense take those risk factors inside our system in the form of tremendous complexity. The job Mike had told us to go do would prove bigger than any of us could have foreseen.

When I started working EDL for *Mars Science Laboratory,* the idea had been to use two parachutes, a *Viking*-heritage supersonic design to break the initial speed and a second subsonic parachute that we would custom make. From the get-go I had felt that relying on a *Viking* replica design to slow a spacecraft that was nothing like *Viking* was nuts. And I was not alone in this. Bob Mitcheltree, for instance, a brilliant Ph.D. engineer who had come to JPL from NASA Langley, agreed that the same *Viking* parachute design and size was too small to use with our bigger spacecraft. He had been working on a bigger subsonic parachute that we would employ after the supersonic chute had slowed us down. He had spent millions of dollars on testing, but he still didn't think we had the right supersonic-parachute sizing.

Even so, this was another case in which respecting the human meant giving people enough time to come to terms when their ideas had reached the end of the line.

A lot of really good engineers had invested heavily in the idea of two parachutes, so we held a couple of design team meetings during which I made the basic argument that the two-parachute idea was not the right solution. The all-important correlation that had been missed was the relationship between the vehicle size and the parachute size. Everybody in the room could follow my logic and see where this was going, but I never pulled the trigger to just come out and say, "This two-parachute idea is dead." Instead we scheduled a decision meeting for a couple of weeks out. The writing was on the wall, but it gave people time to get their heads around the change.

The single, huge parachute we then went on to commission, however, was so big that it raised other concerns.

Viking's chutes had been 16.1 meters (52.8 feet) in diameter. We were looking at 21.5 meters (70.5 feet), an expansion of roughly 15 percent. Some of our colleagues on our EDL review board worried that this increase in scale was enough to change the basic physics. The governing equations of fluid dynamics suggested otherwise, but we couldn't know with absolute certainty, because we'd never been able to do a truly valid test.

We were planning to open our one big chute at Mach 2.25, more than twice the speed of sound. If you were to open a chute at speeds that fast within Earth's atmosphere, the material would instantly rip apart. Temperature and atmospheric density, both very different on Mars, have a powerful effect on aerodynamic performance. The only way you could mirror Martian conditions would be to release your test parachute at an altitude of 130,000 feet or more while flying at twice the speed of sound. In addition, to account for the wake of the spacecraft, the object releasing the chute would have to be of similar size and shape. All of which made atmospheric testing on Earth, if not impossible, then prohibitively expensive.

Pathfinder had used a slightly smaller, modified version of a *Viking* parachute. It had been modified to lessen the swinging under terminal-descent conditions; the modification had seemed to help, but the evidence was not crystal clear.

For MER I'd advocated growing the parachute a bit, and when I'd examined the process of earlier testing during *Pathfinder,* I'd come to the conclusion that they never really achieved the level of precision and certainty they were looking for.

During the 1970s the *Viking* mission had spent more than

$100 million on an extensive series of tests, both supersonic and subsonic, to qualify parachutes. They took high-speed film of these chutes performing and, as a result, indirectly established what came to be known as the *Viking* box, a portion of a graph with dynamic pressure on one axis and Mach number on the other, defining the parameters thought to be safe for parachute deployment. Since the late '60s, engineers had treated the region of Mach number and dynamic pressure space outside the range *Viking* had tested as an uncharted territory where "there be dragons."

The "paper study" guys especially liked to pull out the *Viking* box in every discussion, but the model was more precise than it was accurate, which can easily poison perception. I hated the *Viking* box. I actually forbade anyone from ever making a plot that had the *Viking* box on it. I wanted us to look at *all* the test data, not just the *Viking* tests. I did not want us to accept the parameter of the box as the safe zone while tagging anything outside the box as the death zone.

When we analyze issues pertaining to EDL, one of the key tools we use is Monte Carlo analysis, which means that you do thousands of runs and see how the physics of the various random processes changes our trajectory results. In the case of the parachute, these analyses give you a range of your opening conditions. That way you get to the whole truth of the physics, and then you draw your conclusions based on engineering judgment.

We wanted to plot our "cloud" of the thousand opening conditions we might expect and compare them to all the test data we had, not just the *Viking* test conditions. With all the details of all the tests, you could talk about actual results rather than accepting a derived model of the results in the *Viking* box said to represent a safe zone. The *Viking* box model of the world was oversimplified, and to be frank, with the size of our rover, we were operating outside the box anyway.

We pulled out all the tests from *Viking* and from others and showed all the spreads and aligned them with our parachute deployment cloud to establish a real sense of the risks. We began to develop the argument, grounded in basic physics and logical inferences, that the inflated shape is basically the same at both supersonic and subsonic speeds and that, as a result, the stresses would be the same. While the old *Viking* box wasn't really going to help us, the fundamentals might.

Helping me pitch this concept to our review board was Juan Cruz, the Langley aerodynamicist who'd come to Boise to help with our tests for MER. Originally from Puerto Rico, Juan wore his thick, wiry red hair in a mullet that fell halfway down his back. He was also a classic engineer, at least in the sense that he liked to lay all the data on the table, without using the kind of selection and shaping that engineers often attribute to "marketing people" and deride as being slick.

We made the case on basic physics, and we were supported by all the old test data, not just the data from the *Viking* box. The *Viking*-heritage formula had a ratio of parachute diameter to spacecraft diameter that could be scaled, and the other tests had been done at higher Mach numbers and under different pressures. Which should be enough to convince anyone.

In front of the review board, I let Juan do most of the talking, and by lunchtime everyone seemed to be with us. But Juan had brought along several of the old *Viking* test videos, and now he wanted to show them.

The videos of the tests that had created the *Viking* box parameters had been lost. The ones Juan had were precursor videos that constituted more of a freak show of occurrences outside the boundaries. Some showed parachutes opening at Mach 3+ and melting in the attendant friction. None of these images had anything to do with the conditions we would face or the case we were trying to make. They were also terrifying, and for that reason alone, they should have been left at home.

I'd gone over all Juan's slides, but for some reason I'd let him assemble the videos without close examination. Certainly I had not thought to discuss their subliminal message or the kind of impact they were going to have, which was to leave the board members pissing in their pants with fear.

After lunch but before the screening, the board had exuded a nice warm feeling of confidence. After the screening they demanded that we order more tests to bolster our argument. They had just seen in graphic detail parachutes falling apart in high-altitude supersonic tests, whether or not those failures had any relevance to the point we were trying to make or to our proposed use for the parachute. These were graphic images of conditions wholly apart from those we faced. We understood the test outcomes, and we knew that the phenomena that caused the failures were not anything we needed to worry about.

Showing these irrelevant and terrifying videos was a classic case of "selling past the close" and losing the deal as a result. Technical people sometimes have trouble with this, but you can't get away with simply offering unfiltered information.

Speaking the truth, the whole truth, and nothing but the truth is not enough. You also have to take responsibility for understanding what you're saying, why you're saying it, and the effect that hearing it is going to have on your listener. The truth is not just the collection of all the facts you may know in relation to a given subject. The truth means understanding the significance of those facts, the validity, the true representative nature of each of them, and being responsible for which ones should be shared in order to effectively *convey,* not just state, the truth.

Vomiting up all the facts onto the table has a way of clouding truth. Similarly, saying too much without helping the listener understand what is important and what is not is irresponsible. To really convey the point, one must prioritize the facts, sort the wheat from the chaff. Admittedly this selective editing process puts the editor in a dangerous position. It means that if you are not clear about your own motives and your ultimate intent, you can cloud and color the truth in ways that are not just deceptive or manipulative but destructive.

When I'm advising engineers how to explain things, I often say, "Sometimes to express the essence of a thing, you have to be willing to do violence to the fact of it." This statement is supposed to arouse feelings of distress and anxiety. Yes, "Is he telling me to lie?" is the question I am trying to elicit. I most certainly am not, but I do want people to feel as culpable as if they were telling a lie. I want them to understand that to really communicate, you have to take personal responsibility not just for the validity of the facts but for conveying the underlying truth that they represent.

Ultimately, because of our unfiltered, unsorted truth, we would have to spend another two years and $10 million trying to prove that we were right but mostly trying to erase the emotional distress of those confusing images.

The fundamental problem we had to answer was this: Does the

same physics apply when operating in the same relative length scales as *Viking* but at larger actual sizes? In other words, do the laws of physics, as applied to parachutes, scale?

The basic governing laws say that the physics should scale and there should be no surprises. The review board, however, wanted hard analysis or test data. Trouble is, the problem lay well beyond the state of the art in computational fluid dynamics, so our added challenge was to sneak up and reinforce our understanding of this problem without having to make breakthroughs in analytic technique so far ahead of the field that they would be worthy of a half-dozen dissertations.

We went back to the wind tunnels at Glenn and Ames, and we produced subscale supersonic tests of rigid parachute models trailing behind bodies representing the size and shape of our spacecraft. We were able to visualize the flow and see the phenomena of the interaction of the wake of the spacecraft with the pressurized regions of the parachutes and how that flow would pressurize and depressurize the parachute. In the case of a flexible parachute, this flow would lead to its collapse and reinflation, or perhaps just to collapse, or perhaps we would see the process of collapse and reinflation tearing the parachute apart. There were few bounds to the places our fears could take us.

When you're going at supersonic speeds, you have the airstream going over the spacecraft and feeding the parachute. But as the air comes around, the spacecraft slows it down, even stops some of it, and the air gets mixed up, and so you have this small region directly behind the spacecraft and extending toward the parachute that's going slower than the speed of sound, creating a subsonic wake, a region in which the flow is moving slower than the speed of sound.

So when our capsule goes through the atmosphere of Mars, it's pressuring the air in front of it, but if it's moving faster than the speed of sound, those pressure waves can't escape, and they get balled up into a shock wave. When you're moving faster than the speed of sound, it also means that no information can move upstream, so the spacecraft has a shock wave and the parachute has a shock wave. Information and flow changes from the parachute can't move forward and interact directly with the flow capsule.

But in that region of slower-moving, subsonic air directly behind the spacecraft—the subsonic wake—information, including pressure, can flow up and down. So the high pressure from the parachute can squeeze up this tube of subsonic air and depressurize the parachute—until the pressure is reduced, the wake collapses, the parachute reinflates, and then it happens again.

In the wind tunnel, we used rigid models to gain an understanding of these fundamental physics. If you open up extremely fast (above Mach 2.4), this wake deficit goes on only so long behind the spacecraft before the slower air eventually recovers supersonic flow. That is, your subsonic wake region becomes shorter the faster you go. Eventually, if you go fast enough (again above about Mach 2.4), the subsonic wake never makes it back to the parachute bow shock and the two bodies are once again aerodynamically decoupled. You can no longer collapse the parachute.

Testing during the *Viking* era had shown this phenomenon above Mach 1.5, and our small-scale wind-tunnel tests had the same behavior, turning on at Mach 1.5 and turning off again at Mach 2.4. So all this testing mostly just confirmed subscale observations that had already been made at full scale years earlier. Our efforts simply allowed us to say with more confidence what we'd said at the beginning: that the physics holds regardless of increased size.

To me, this whole experience of selling past the close and having to invest two more years in inconclusive testing reinforced two classic lawyering lessons about the control of information: You never ask a question you don't know the answer to, and the battle is won or lost over which piece of evidence the jury is allowed to see.

The next issue we had to resolve was radar. We had chosen to build our own, with five pencil beams, each a very narrow 3 degrees in width. We liked the narrow beams for many reasons, primarily because they interacted well with the Mars terrain. A wide beam gives a strange average of distances below, which means that a lumpy surface can give you deceptive readings.

Looking straight down on a hill, a wide beam gives you a lot of radar return signal of the ground around the hill. Get closer and the wide beam starts to illuminate more of the top of the hill. All of a sudden, when you thought you were hundreds of meters above the terrain, you find that you're maybe only a dozen. Which can be, of course, incredibly dangerous.

Another reason we liked the idea of narrow beams was that they make it easier to place the antennas. Our spacecraft was getting crowded, and as we looked to fit a radar on the descent stage, above the rover but underneath the back shell, we were having a hard time fitting stuff and finding a place for our radar antennae to point. The radar antennae, which looked like small dinner plates, "shined" their beams kind of like flashlights, only the beams were invisible to the human eye. We assumed we could get the skinny beams to shine between the wheels of our rover as those wheels protruded below the back shell on its descent toward Mars. But then we realized that in the near field, even these pencil beams aren't skinny at the beginning. Up close the beams are very wide, which means that if they were to be shining between the wheels, they were going to see the wheels, with the beam perhaps reflecting off them and back into the antenna. This would spoof our radar. It is like holding your hand in front of a flashlight.

We were forced to rethink. The solution to this problem of reflective interference was for the wheels to remain tucked up against the rover as long as possible before contact with the surface, and for the rover's mobility to be deployed in Sky Crane maneuver as late as possible, just 20 meters (65.6 feet) above the ground. We also had to create two special antennae pointed away from *Curiosity* as it descended in the Sky Crane maneuver. This meant four antennae for the initial descent and two during Sky Crane.

Our original plan had been to deploy the wheels during the portion of our EDL sequence while we were suspended under the parachute. There we'd have a leisurely minute or so in which to complete the task. But now we were piling in more and more activities that had to be conducted during the Sky Crane maneuver, which meant during the last few seconds of EDL. And every change costs money, and every change carries potential risk.

What we came up with for the last-minute wheel deployment was to keep it fast by simply relying on gravity. But would the mobility system, with its wheels, be strong enough to survive that kind of quick drop-and-jerk into place?

The problem was what's called sensed acceleration, meaning that the apparent weight of the rover depends on the functioning of Sky Crane's lowering winch, called the BUD (bridle, umbilical, descent-rate limiter). The BUD is a spool attached to an electric motor attached to a bunch of resistors. Gravity pulls the cable, which spins the motor, which creates resistance. If we let it all go at once, the wheels would be weightless, the same way you feel weightless in the first instant when an old-fashioned elevator begins to drop.

But the fact that our motive force for the wheel deployment would be the output of another motive force—the performance of the winch in resisting gravity—compounded the uncertainty of how the rover was going to react to the quick drop-and-jerk.

Chris Voorhees, the mechanical systems lead for the rover, noted that one of our standard requirements was that all of our EDL deployments in space had to be tested on Earth. Given that Earth's gravity is much stronger than Mars's gravity, if we tested the mobility/wheel drop on Earth and everything fell into place without a problem, then the system would necessarily be strong enough to work on Mars, regardless of how the BUD was tuned or its uncertainties.

But as things frequently happen, Chris got busy with something else—an issue with the sampling system, I believe—and he handed the mobility system to another engineer on the rover team who did not fully appreciate the criticality of wheel strength for this aspect of our landing and did not fully understand our EDL test-on-Earth requirement. So without talking to the landing team, she asked for a waiver on the test. Oblivious to the implications, she simply called up a trajectory analyst and asked, "What loading do you see in the mobility system?"

By the time I got wind of this, it was six months later and many pieces of the wheels and the suspension system had been designed and/or fabricated based on a loading estimate that came not from a test on Earth but out of a trajectory simulation. Although this estimate was valid, it was not constructed to bound all the possible loading we could encounter in flight, specifically the drop-and-jerk. Which left us in an extremely vulnerable position.

The result: Very late in the game, we had to do a load of complex nonlinear stress analysis and testing to failure in order to prove that our landing system, which ultimately relied on the rover wheels, would work. When we were done, we had convinced ourselves that we were in okay shape but not in as robust a position as we had initially planned.

We got into this situation not only because a question was framed the wrong way but because it was asked and answered by people who didn't have the necessary insight into how the information would be used. They also didn't have the visceral connection to the problem at hand that would contribute to the necessary insight. They were thinking wheels rolling on the surface of Mars. They were not thinking wheels and wheel struts sustaining the impact of a drop-and-jerk.

Failure of communication like this is a fairly classic problem that occurs in any organization. The essential, critical information, even if written down (which we had also failed to do in this case), is often forgotten in later decision making when that occurs at an organizational or temporal distance far from when and where the critical essence was identified.

The waiving of the test-on-Earth requirement had taken place at a meeting where a presence from the landing team was not thought to be required. The larger problem was that the mobility team, the wheels guys, identified themselves as part of the surface mission, meaning the rover, rather than EDL. They knew there was an EDL component to their work, but they had never been made to feel that they were a part of it. This created a separation between those who understood the importance of the test-on-Earth requirement and those living with (or asking to live without) it. The essential had been lost in the decision making.

Some people might hear this story and blame a lack of documentation. They're right, but that's not the lesson I take away. I believe that the only way for a group to really hold on to the essence of the task at hand is for all the members of the team to feel deeply intellectually and emotionally invested in the same vision. A more cohesive team would have sensed the disconnect. The lesson I learned from this mistake was that you not only have to find something to love about each of your people; you have to extend that love far and wide to all the other divisions and departments that you depend on.

I should have said to the wheel team in the rover design group, "You're our landing gear, so you've got to be in our meetings with us, helping us solve our problems." And they should have extended that same welcome to me. That's the dream for systems integration. A more cohesive team exchanges information more efficiently and has a stake in the efforts of each of the elements within the team as a whole. The members look out for one another better and, in doing so, look out for the overarching interest of the project.

The problem also underscores the fact that there's more to communication than asking the seemingly appropriate question and getting a seemingly appropriate answer. You have to get down to that deeper truth I keep talking about, the essence, which has to be shared within context and with connectivity, without which you're vulnerable to all sorts of errors that will look oh-so-avoidable after the fact.

Again, it was my job to make sure that everyone was communicating with the right degree of engagement, and if they weren't, it was my fault. That said, we also faced a structural challenge. Our communication was not helped by the fact that the mechanical engineering effort was divided, with a manager for the rovers and a manager for launch, cruise, and EDL. The folks in charge of rover mobility reported to a manager who didn't have Entry, Descent, and Landing as part of his explicit task. As a result, they spent all their time worrying about how the wheels were going to work on Mars and not about how they would function en route.

Our salvation in all this is that we usually have more strength in our structures than we can take credit for. When you go to break something made of a metal, it usually bends and deforms before it breaks outright. In fact, usually we consider a part "broken" when it has bent the slightest bit, almost imperceptible to the naked eye. Ultimately we convinced ourselves that these parts could still work even if slightly bent. We tortured the mobility system with a hydraulic ram to replicate the drop-and-jerk into place, and it all hung together, actually bending the system the way we thought it might in a worst case in flight, and we concluded that we were going to be okay. In doing this, we essentially cashed in all of our reserve strength and margin and committed to flying with our asses hanging out.

Sometimes when you're struggling with a development, issues just keep popping up, one after another, like blows on a boxer against the ropes. So it was at this time for us on MSL. Not only were we tackling how to provide lift, to trim tab or not to trim tab; whether our world's largest supersonic parachute would actually work; and how to deploy our wheels safely prior to landing; but the sheer size of our vehicle meant that we were going to heat up more than ever before when we hit the atmosphere of Mars going more than 13,000 miles per hour.

Mars Science Laboratory was going to be the largest blunt-body aeroshell we'd ever pushed into any atmosphere for any mission in history. Shuttles were bigger, but the higher up you go, the greater your speed as you're coming down. Coming down from low Earth orbit, the shuttle never traveled nearly as fast as MSL was going to be going. *Apollo* came from the moon very fast, but it was smaller.

One of several technical challenges you face when you're bringing in something very big very fast is that the way the atmosphere flowing across the heat shield changes depends on the size of the shield. Turbulence increases with size, mixing the hot gases that are far from the heat shield down into closer proximity with the thermal protection material.

Given MSL's size, we knew that our thermal protection challenge was going to be more intense than on previous Mars missions. But I still wanted us to leverage the thermal protection system (TPS)—using SLA-561V, a kind of rubbery, cork-based material—that had worked well at heating conditions encountered on *Viking, Pathfinder,* and MER. I was also skeptical about the dark predictions of just how bad our heating environment was going to be.

The process of analyzing any engineering system as complex as the aerothermal environment is filled with uncertainty. Upon each encounter with uncertainty, a given engineer will choose the more conservative path, meaning that he or she will assume the worst-case scenario. Then the engineer who comes next does the same thing and also assumes the worst. So I suspected that our aerothermal environment wasn't going to be quite as bad as it was assumed to be, and I had hopes of being able to persuade some of our team's aerothermal-domain experts to let go of the conservatism that had been baked in. I had pushed them hard to accept flying the tried-and-true SLA-561V as our heat shield TPS, even though the heating conditions were much higher than before, at least on paper.

This all came to a head at the time of our critical design review—CDR in project-speak—at JPL's Theodore von Kármán Auditorium in June 2007. This is the review at which you're supposed to know that everything is going to work and you're just torturing the details of why it will work.

I stood on a dais on a stage in front of the project standing review board and a couple of hundred onlookers and went through my slides, fielding questions. I made the case that our TPS material met our needs and that we'd tested up to the edge of the worst case and it had done fine.

Just as I was wrapping up, the cell phone in my pocket started to buzz. Somebody was calling again and again. It was distracting, and it kind of pissed me off. Everyone in the team knew where I was at that moment.

I finished the presentation and left the podium, then went outside to see who'd been so eager to reach me. It was James Reuther from the Ames Research Center in the San Francisco Bay Area, a member of my EDL review board who is a specialist in aerothermal analysis and TPS. He was also in charge of the advance development of the heat shield for a new joint human-robotic mission to the moon called *Orion*. They had been working with SLA-561V, testing its capacity to bring back astronauts from low Earth orbit, but low Earth orbit did not present conditions as extreme as ours. They had exposed SLA-561V to the midrange of temperatures, he told me, and even so, the material—the same material I was advocating for us to use—had simply evaporated.

This was not good news.

"What did these samples actually do in the test?" I asked.

"We don't know. One moment the material was there, working well, creating a char layer with very little melt flow, and then it just started to disappear," James said. "This was at an intermediate condition. Not even worst-case heating."

I searched for a reason to stay calm, even though I could feel the edges of panic creep up as I paced back and forth in front of the auditorium where I had just told the project standing review board that everything was fine with our TPS material.

SLA-561V is packed into the holes of a honeycomb substrate—standard procedure. But James said the *Orion* team had wanted the material thinner, so they'd cut it and then used the two sides. This meant that their conditions were slightly different. Maybe they'd been off just a little. At least I could hope. But I also had to test.

We took a batch of fresh SLA-561V, put together by our team to our specs, to an arc jet at a different facility, Arnold Engineering Development Complex in Ohio. (An arc jet is a wind tunnel where we can blow a column of air the size of a pizza through a huge set of electric arcs, creating a superheated jet of gas to replicate the high aerodynamic frictional environment that a heat shield encounters as it enters an atmosphere.) We heated it up, and the same thing happened. The same material that did well at extremes simply disappeared under moderate aerothermal conditions. We conducted test after test and watched it evaporate every time.

We decided to conduct one last check to attempt to reproduce the results that *Orion* had seen back at Ames, using our own SLA material. Of course, this, too, was likely to be a repeat of the *Orion* data, but we had to double- and triple-check this, because if it was true, the news was really, really bad for us.

I went up to the Ames Arc Jet Complex to see these tests myself. In the tunnel there was a small quartz window where I could look in with special glasses. I peered into the window and watched our TPS sample heat up and glow white-hot (normal); watched the top surface start to melt a bit, get soft, and flow a bit (normal) and then, as if some evil spell had been cast, suddenly disappear. Just disappear. The SLA-561V just seemed to vanish from inside the honeycomb cell. If that were to happen on Mars, it would be game over. I was staring mission death in the face through that quartz window.

After walking out of the arc jet, a bit stunned, I phoned Richard and Matt and said, "Hey, guys. We have to change our thermal protection material."

We were now two years from our scheduled launch date, and we still didn't know what we were going to use to protect our spacecraft from incineration. So we started to scramble, running various options in parallel, and I brought in a frequent go-to guy, a specializing nonspecialist named Eric Slimko, and we laid out a matrix of all the possible materials we could use.

As we went through the list, we knew that we had very limited time for research, so part of what we were considering was "What do we have the most information about?"

Engineering, as I've said before, is often not about being right, as in perfectly right, but being right enough. Frequently, when pushed for time, you do not have the luxury of looking for the absolute best solution. In situations like these, you have to balance performance with the greater certainty that comes from a bird in the hand. In this case, the bird in hand was PICA.

PICA (phenolic impregnated carbon ablator) had been used for a smaller heat shield for *Genesis,* a sample-and-return mission launched in 2001 to study solar wind. It usually came as a single molded unit about the size of a two-top diner table. But because MSL was so big and no one could make PICA in a single unit big enough, we were going to need to use it as tiles.

The *Orion* team, developing their joint human-robotic mission to the moon, had looked at tiled PICA and were sufficiently concerned about the caulking between the tiles, along with the interaction of the tiles, that they were going to pass. Still, the material itself had great promise.

We had a good amount of data on PICA, and because it was very stout, it was not likely to exhibit the kind of complex behavior that SLA-561V did at those intermediate conditions. And unlike SLA, it was made up of two things, not dozens of things. This attracted us, since SLA had surprised us by failing not at the extremes but in the middle of the possible aerothermal conditions. Such behavior is what engineers like to call nonlinear, meaning it does not show a simple one-to-one relationship. We wanted a TPS solution so simple that it wouldn't surprise us. The downside was that PICA was more capable than we really needed—at the cost of being really heavy. But it fit the bill in terms of predictability, and that was our biggest need at the moment.

So I got on a plane and flew to Plymouth, Maine, with several colleagues to visit Fiber Materials Inc., the people who make PICA, to learn about the risk associated with making it into tiles. If we went ahead, we were going to need two roomfuls of these PICA tiles, and we were going to need them day after tomorrow. The company convinced us that tiles could solve our problem.

We got our PICA, we machined the tiles, and we put the capable folks at Lockheed Martin to work integrating them into the heat shield. Normally such an effort to design a new TPS system from scratch would take eighteen to twenty-four months. We did it in nine. It is always surprising how much you can get done when you really need to.

The lesson, if there is one, is that no one's engineering judgment is foolproof. I believe in a determinant universe, one that (a) exists and (b) is ordered by laws that cannot be changed at will. The search for the truth about such a universe invariably involves building a model of all that we know, which is an adjacent or surrogate universe.

I had been making my judgment about the TPS based on my model of the known universe. In my model the worst-case conditions were found at the edge, or the extreme, of the environment. This is a natural supposition. If heat is bad, more heat is worse, and the most heat is the absolute worst. But it turns out that my natural assumption was wrong. Just as my mom taught me, when you assume, you make an ass out of you and me.

A careful balance needs to be struck between our confidence in the parts of the universe we know and our suspicion of the universe we don't know. Drawing on Rumsfeldian parlance once again—your known knowns, your known unknowns, and your unknown unknowns, or "unk-unks"—I think the art of engineering, or being successful in business or life, for that matter, lies in how one finds the unk-unks and makes them "kn-unks," and how one respects the possibility for more to be found . . . or never discovered.

In my experience, the older and/or wiser one gets, the better one is at judging the unk-unks, sensing them, getting a feel for how vast they may or may not be, and somehow managing them. This is all done, of course, without actual conscious, explicit description of or actual counting of them, since doing so would immediately thrust them into the kn-unk category. I've known many smart people, those guys who won the physics prizes back in high school and aced the exams at Caltech, who tend to underestimate the unk-unks. They are aware of their intelligence, and their model of the universe is one in which that intelligence will always prevail. I've been in rocket science for quite a while and endured its humbling tutelage, and yet at times I still find myself in their number.

Recently I've been working to help develop a sampling system for the next big expedition to Mars. This system will extract core samples of rocky material from the Martian surface and seal them in special containers for priority shipment to Earth. One of our main challenges is keeping all of our equipment very, very clean, so that when we bring the samples into our world and unseal them, we don't find Earth-born bacteria or other life forms that we put there and conclude, wrongly, that Mars is alive.

When you look closely at a superclean spacecraft being assembled in a special clean room by technicians in head-to-toe white "bunny suits," you still find all sorts of microorganisms hanging out, waiting for a ride. So the challenge is big enough without the fact that biologists and physicists don't speak the same language. For biologists, microbiotic life is everywhere and very hard (if not impossible) to track. For physicists, microbes are just like any other particles bouncing and being blown through the air, and we should be able to track their motions. Bridging the gap requires that everyone recognize that each school is relying on a model of the universe that is not the universe per se. It's only when we recognize that our "truth" is only a model, and when we let our models compete with other models, that we get anywhere.

Chapter 11

APPRECIATING THE MUNDANE

WHEN YOU'RE BUILDING SOMETHING LIKE AN ENTRY, DESCENT, and Landing system to put a spacecraft on Mars, it's easy to devote all your attention to the "naughty bits," things like phenolic impregnated carbon ablators, Mach 2.25 disk-gap-band parachutes, and high-thrust, deep-throttling hydrazine monopropellant engines, the innovative, glamorous new designs that push the boundaries of the physical laws. And it's all too easy to completely ignore the more mundane elements that are just as essential. It pays to remember that you can have written the most sophisticated EDL software known to man, but all it takes to kill your billion-dollar project is failure in one component the likes of which you could have picked up at RadioShack.

This is not easy for me, personally. I don't like it. I am frustrated by the fact that our new and novel system can be brought down by functional parts that our state-of-the-art system shares with every other spacecraft launched since the sixties.

When I took on EDL, my first priority was to solidify the architecture and the team. The next task that engaged my attention was to scrub the shit out of everything we'd put together, to really understand and perfect every aspect of every path we'd chosen. We'd spent years on the cutting edge, bearing down on the naughty bits like Sky Crane and our radar system, but now we needed to focus on the stuff that had nothing intrinsically to do with EDL. These were the mundane components that were easily budgeted and predictable, like radios for deep-space communication, batteries for our power system, celestial sensors that were copies of ones we had used for decades, flight software, and so on. All they were asked to do was to execute modest algorithms, to talk to one another, to take the right measurement, to have their clocks interact. The catch was that they needed to execute these algorithms absolutely without fail.

There's a similar challenge in appreciating the seemingly mundane when it comes to people. It's all too easy to take for granted the workers who don't spend their time exploring the theoretical frontier, the ones who are not flashy or flamboyant, who are not talkers, who have no interest in being like the sun shining on everyone, but who are nonetheless exceptionally good at tightening the bolts, checking the data registers, and making stuff work.

Not only do you need both types, but every type on every team needs sufficient self-awareness to know how the larger interaction is unfolding and how everyone's job—no matter how solitary or relatively unglamorous—fits into the larger scheme of things.

In EDL for MSL, we had a bottleneck in the person of an engineer whom I will call "Michael." The number one sin that could get you on my shit list is failure to respect and speak the truth. The number two sin is withholding information. Michael was a withholder.

My experience with him was my foremost failure as a manager of people.

Michael was already on the team in 2005 when I stepped in. He sat in the test bed, a roomful of computers that are versions of what we're going to have on the spacecraft, and his specialty was the interface between the housekeeping functions of the software and the razzmatazz of EDL. In other words, he was in charge of all the basic utility functions that could just as easily serve a municipal power grid or a nuclear reactor, systems that were vitally important but had nothing inherently to do with traveling through space and landing on Mars. He was working on the stuff I had a hard time caring about, or, to tell the truth, respecting.

Guidance, Navigation, and Control—the automated behavior that flies the spacecraft—is one of the naughty bits, but below that is software that actually carries out the GNC's high-minded intentions. A command like "Open the parachute" may require a series of six distinct behaviors. For me, this is a level of detail equivalent to what goes on inside my smartphone. My phone can do all sorts of amazing things, but I don't really give a flip *how* it does them. All I want is that, when I push the button, the call goes through. The same is true on the spacecraft. When we send the command to close the antenna, I feel that the antenna should close—done. I do understand that this perspective is perhaps immature to the point of being infantile, but here's my frustration: How is it that something we have had to do hundreds or thousands of times in the past can still not be a given? One hundred years ago, successfully starting a newfangled automobile was perhaps an eight-times-out-of-ten proposition. Today a failure of your car to start when you twist the key (or push a button) is more like a one-in-ten-thousand event. We can now assume that the basic stuff will work.

Even though we (humans, Americans, JPLers) have been building spacecraft for six decades or so, the economies of scale and production mean we haven't had the chance to perfect reliability, as we have with an automobile or a smartphone. It's frustrating and annoying, but it's true.

Michael worked with the systems that are supposed to do these functions. Part programmer, tester, mapper of functions, he worked with the software and the subsystems that made those kinds of things happen. And he was very good at it.

Michael had a keen but perhaps fragile sense of ownership over his work. My challenge with him was that this sense of ownership translated into overly proprietary behavior. He wouldn't teach, and he wouldn't share—not with me and not with anyone I sent his way. In retrospect I think he felt that his power emanated from the fact that nobody else understood his job; if he opened up about it, he might lose his exclusive hold.

Part of my job was to be certain that the bits under Michael's domain were functioning smoothly and that they were properly integrated with what everyone else was doing, mundane and naughty alike. Unfortunately Michael seemed to regard my interest in his work as a threat.

Eventually I segued from soft nudging to more direct approaches. Questions like "How are the interactions with power subsystem going?" turned into "Can you teach me how fault containment in the high-side/low-side switching works?"

In late 2006 I hired a smart young guy named Steve Sell who had broad experience similar to Michael's. I knew he would talk to me, and so I sent him down to the test bed. "Go get inside this so I can learn it," I said.

Michael's immediate response was "I'm quitting the project."

I should have let him. My Spidey sense said, "Cut it off and let it bleed." But I didn't listen. Although I didn't like the emotional blackmail of "I'm quitting," I worried that we wouldn't have time to adjust to Michael's departure. Of course, you always can do more than you think, and we probably would have been okay if he had left. But I didn't have the courage to face my senior advisers and mentors, all of whom were telling me that it was my personal approach, my combination of in-your-face style and insufficient love of Michael's subject matter—not Michael's attitude—that was the problem. So he stayed, and matters went from bad to worse.

When we started out on *Curiosity* in 2003, our initial target for launch was 2009. We were on a roll after MER, and success fuels further ambition, but we needed to temper that ambition with the right practical caution. And caution gets most challenging in times of success. How many companies have failed with their second major product rollout, not long after the first stunning victory?

Looking toward our fast-approaching launch date, we in EDL were nearer the ready-to-go end of the spectrum than most of our colleagues. For instance, large chunks of the flight electronics being built and optimized for things other than EDL had to be integrated, and these elements could hold us back. We would all be using the same onboard computer, for instance, but we couldn't get into writing our EDL software, let alone testing it, because that computer wasn't up yet.

The project also had overreached with a few too many exotic solutions that led to hang-ups. Most notably, we'd tried to create lightweight and cold-resistant gearboxes that sat at the end of the many motors distributed through the spacecraft, including on the wheels. These were made of light titanium and used a special hard-coated powder rather than grease to create a dry lubricant that could function in super cold environments. Unfortunately these processes did not converge, and the gearboxes kept failing in their life tests, and that ate up far more time than it should have.

The people who build spacecraft to explore the solar system can do this job only if they don't freak out when they're in the Dark Room, that place where you know you have no solution to your problem. So you develop resilience in order to avoid panic. But the effort to be resilient and keep marching forward can sometimes mean that you don't take reparative action when things really are as bad as they seem. You're working hard to stay in calm mode when occasionally it's appropriate to freak out and hit the alarm.

By 2007 it *should have* been obvious to all of us that we weren't going to make our deadlines for the 2009 launch, but we were under pressure and we were committed, so we bought into a kind of collective tunnel vision, kept our heads down, and told ourselves that we were going to make it if we just kept pressing on.

The fact was, we were simply too close to the problem, too invested, too wrapped up in it. We needed more time to step back, get our minds off the high-velocity treadmill, and get down to the deeper truth of the issues that were holding us back.

What we needed was more space for the integration of conscious and subconscious. But what we also needed was a doubter, someone from the outside with a beginner's mind, who could look at our progress and call out the truth, which happened to be "You need to rethink on a major scale."

The man to fill that role was Rob Manning, the *Pathfinder* guy who had helped us prepare our presentation for Mike Griffin at NASA. He had been working on a project called *Phoenix*, which would put a lander in the Mars polar region. He also had a long history with Richard Cook, project manager of MSL, on the previous

Mars missions, and so there was a lot of trust when he came to us as chief engineer in 2007.

Having joined the project late, Rob stood outside the tunnel the rest of us were in, which allowed him to see the truth. He also had the wherewithal to convince Richard of what needed to be done, even when the truth called for drastic measures.

When Rob came on, he started walking around and saying to people, "Tell me, what's your worst nightmare?" He collected all these worries into what became known as the Manning list. Worrying and finding problem areas were also just part of Rob's personality, and before long he started working on Richard to make the case that we were not going to be ready for our launch date.

Rob was incredibly persistent, and by the fall of 2008, JPL officially acknowledged what became known as The Slip, and the launch date for MSL was moved back from 2009 to 2011.

Whenever there's a really big problem, you need fresh blood and fresh ideas to make sure you're covering all the bases. That said, there's also an element of theater that comes into play. You can't announce a major error, then not do anything to rectify the situation.

The most significant change as a result of The Slip was that Pete Theisinger came back from a higher administrative function to take the helm directly, and Richard Cook assumed the role of deputy project manager. This was, of course, a blow to Richard, who, whether he deserved it or not, was the one ass to kick. And a kick he received.

But outside help also began to pour in. Dara Sabahi contributed from the vantage point of his new position running something called the Integrated System Engineering Office, a relatively short-lived, independent fiefdom that was in some ways more like a shadow government. When he examined EDL, one of the things he started beating me up about was my handling of Michael. "You're scaring him," he said. "Not being sensitive enough. He just needs the right boss."

Dara took over and essentially took him out from under my leadership. Michael now had direct access to Dara, but Michael didn't want to talk with Dara, either, or to teach him, or to connect in any way. Eventually Dara gave Michael over to his chief lieutenant, Joel Krajewski, a strategy that also didn't work.

Eventually Michael was given back to me, but because of all the data that he had been keeping bottled up, our information flow and our capacity to test had fallen even further behind.

According to JPL's matrix organization, you have a supervisor for the project you're working on and an administrative boss who looks after you as a human being. Michael's administrative boss was Richard Kornfeld, who protected him from my aggressive management style back when I was trying to get Steve Sell into his domain. They had a relationship that I thought I could leverage, so I brought Richard over as the verification and validation lead, made Michael report to him, and gave Richard the job of getting Michael to open up.

Within a couple of months, Richard was struggling, and verification was lagging. In a collaborative effort, new stuff is always coming up that you have to deal with, so you have to shed some of the existing stuff and give responsibility away, delegating and empowering, which is what Michael would not do. He was sequestering the details, keeping them hostage to his own insecurities. The problem would persist for another couple of years.

At the time of The Slip, in 2008, we were burning a million

dollars a day, and the postponement of the launch added four hundred extra days. With that slack in the schedule, and to save money, we needed to get people off the exclusive MSL clock until we needed them again. So most of us were farmed out to spend part of our time on other projects. I was asked by the human spaceflight crowd at Johnson Space Center to help out in designing their EDL system for bringing humans safely back to Earth after trips to the moon and Mars.

I did my time in Houston, and when I got back to Pasadena, MSL was ramping back up, and Richard Cook said to me, "Okay, you've got your team back. Are you ready to start respecting the mundane?" He knew my bias and that I needed to counter it. But he was also asking the same question of himself, of the project as a whole. It was time for all of us to focus on verification and validation of all the small things. Which began with finding a way to solve the problem of working with Michael.

I had now come to realize that we needed someone with horsepower that was more relevant, more in tune with being down in the engine room, someone who was actually fired up about the kind of work Michael did. We needed a closer, and no one closes like Ann Devereaux.

Ann is a brash, tough-talking southerner who worked her way up building radio systems for spacecraft, dealing with exactly the same kind of interconnections, software, and testing details of spacecraft that were Michael's responsibility.

I'm the kind of person who can remember a few key laws from which one can derive the rest of the world's behavior, but I have a poor capacity for disassociated facts. People like Ann have minds that are more list capable, better at attributing connections between otherwise disjointed items in their memories, remembering facts that are simply jumbled together in the box, with no overarching theory to link them.

Ann took a run at Michael, but even she found it tough going. She was, however, ready, willing, and able to get down and dirty into the details, much more so than any of the rest of us. And because she had an experience base closer to Michael's, she was able to simply take over, putting in seriously long hours all the way to the end. Eventually she hired several young engineers to cover a great deal of Michael's area and substantially wrote him out of the picture, greatly reducing his role compared with what it would have been if he could have shared his work more freely with others. So the very thing that he had worried about for five years actually began to happen, and by his own doing: He was kind of cut out of the loop.

At this point we were beyond the horizon of any new hardware solutions. The basic algorithms of our flight software—how we were going to fly *Curiosity* to the surface—had been in place for six or seven years. But exactly how we coded all that, and how that code interfaced with all the other code to tickle the internal functions of the spacecraft—its power system, radios, on-and-off switches, pyrotechnics—that was the mundane stuff we were only now ready to refine and test.

On June 6, 2011, fifteen members of the EDL team were huddled around consoles in the control room in the spacecraft-handling facility at the Kennedy Space Center. We were six months from launch—a few seconds in spacecraft development time—and we had the actual spacecraft, not a prototype, set up in a clean room, and we were flying the real software. We had a raft of computers sending in information to convince our lander that it was hurtling through the Martian atmosphere on guided entry.

We gave the signal for the rockets to fire, and the valves clicked. There was no mortar deployed in this simulated flight, but from the action of the mechanism, we could confirm that the electrical system had just done what it should to light a mortar had there been one. So we were on the parachute now, and we were about to start looking for the ground with our radar. Then we jettisoned the parachute, and we began our simulated vertical descent down to our simulated Gale Crater, still waiting for the radar to kick in. The spacecraft sat in the other room, searching for a simulated return signal from the ground so the rest of EDL could proceed. We were all hanging there, virtually, gliding down on a parachute heading toward the Martian surface, waiting. And waiting, and waiting. It was taking too long.

Eventually Mig pushed his glasses up on his head. "That's it," he said. "We're dead." He stared at the console. "We're not picking up the ground."

If this had been the real deal, we would have just made a smoking hole on the surface of Mars.

For our crowning test, the most realistic done to date, this was not a pleasing outcome, especially not this close to launch, and it led immediately to a string of all-nighters as we probed every component for the source of the failure. What we discovered was that all our incredibly sophisticated naughty bits had worked fine. There was nothing wrong in our EDL algorithm or our flight software. Nor was there anything wrong with the computer in the spacecraft. What hadn't worked was some run-of-the-mill, everyday components of our test equipment. The test system—the computers that simulated the universe around the spacecraft—had never showed a radar return from the virtual Mars, or at least in the way that the real radar would recognize. Our crowning test ended with us riding our parachute all the way to the virtual surface and hitting it at about 200 miles per hour. Once again it drove home the perils of not giving the mundane its due.

Somehow, during all the stress and strain of MSL, I'd managed to go through a divorce and come out the other side. I'd met a young woman named Trisha Wheeler who worked in the education and communication division at JPL. She was wise to me in a hundred ways, which made her all the more attractive to me.

In August 2011, three months before launch, Trisha and I got married at my uncle Dick's vineyard in Napa, where he had been making wine since 1968. The setting was a beautiful poolside spot at the edge of his vineyard, just below the famous slopes of Stag's Leap. It was a sunset wedding with an outdoor meal, cooked Argentine-style over a large open fire on metal flattop grills called *planchas,* and served on huge tables. My mother was there, but sadly, not my father. The alcohol had finally taken its toll, and he'd died three years earlier.

When I got married the first time, most of those attending were associated with Ruthann or her family. Now I was happy to have my own big family of JPLers, and Miguel San Martin, Tom Rivellini, and others were there, while Richard Cook—thanks to a fluke in California law that allows nonclergy to do such things—actually presided.

There was no time for a big honeymoon. We spent a couple of nights in Palm Springs, then got back to work, moving toward launch. But in relatively short order, Trisha was pregnant.

By October the EDL team had something we could feel comfortable launching. It was not the ultimate representative of what we wanted, but it had a good chance of making it to the surface of Mars. The spacecraft was all bolted together and completely assembled at Cape Canaveral for a final launch-readiness review.

The launch was set for November 25, the day after Thanksgiving. Trisha and I decided to go and to take Caledonia, who had been there to see MER go up as an infant. As Wayne Lee had done on that previous launch, I invited all the EDLers, past and present. Even if my people had to pay their own way, as Wayne's had, at least I could get badges for them.

I flew to Orlando with my family, then drove to the Cape. Cocoa Beach is full of vacation rentals for space people, and we'd booked a condo overlooking the Atlantic and set up housekeeping to make a Thanksgiving meal the day before launch.

We had the traditional turkey dinner for ten, including my uncle Dick, who showed up carrying an imperial magnum (6 liters?!?!) of Steltzner wine slung over his back. He looked so much like my father that I was filled with confused emotions. My dad had forever been stuck inside a prison of his own making, a prison constructed of his fear of failure. I had learned to follow a reckless, daredevil path out of that potential prison—damn the torpedoes!

But over the years, through my time as a student, a teacher, and, most important, a leader of people at JPL, I had also learned to not just close my eyes and jump but to keep my eyes open and have the courage to look where I was leaping. I had learned to hold on to the doubt, to sit in the Dark Room, to know that the worst the universe had to offer was death and that death is not too bad if you do not live in fear of it. I wish that my dad had been there to see this chapter of my life come to pass. I wish he could have shared it with me.

Chapter 12

SEVEN MINUTES OF TERROR

ALL OUR HARDWARE, BOTH NAUGHTY AND MUNDANE, WAS now on board and hurtling through space. But this time, as EDL lead, I wasn't moving on to the next project, and I sure as hell wasn't free to just sit back and watch and worry, with nothing left to obsess over and improve. Software upgrades can be transmitted via radio up until the last minute, so that's where we turned our attention. And with a finite number of hours and minutes, the pressure actually increased.

We also had to refine our operations plans for how the humans would act and respond to data from the spacecraft as our date with destiny approached. We began operational readiness tests almost immediately. If we noticed that it was taking longer than anticipated to generate a data product with an existing team, then we needed to augment the team, which meant that we then needed to practice again with the larger group to perfect their interaction. If we decided to change the ground software to process the data product, we then had to recertify the software.

We also needed a game plan for surprises in the state of the space-craft as it neared Mars. We did tests based on "If we found ourselves this much off, what would we do?" We planned six TCMs (trajectory change maneuvers) during cruise to make sure we hit the right spot at the top of the Martian atmosphere. But because our EDL system used guided entry, it could adjust for an unusual amount of targeting error at the top of the atmosphere, provided that we knew how far off we were. This set up an interesting balancing act (as you may remember from the opening of this book) between cruise risk associated with making trajectory changes and EDL risk.

Meanwhile we continued the verification and validation process, which meant testing, updating, and retesting the software again and again. We beat the crap out of everything, deploying debugging algorithms to scan and search for trouble and trying to invent tests to exercise every line of code. We had people looking at the results of those tests in fantastic detail, which was incredibly painstaking work, like polishing a piece of silver with finer and finer grain cloths, looking for problems that might not even exist.

And here the people who could master the gritty detail that lives inside the mundane were our star players. We had Ann Devereaux and Richard Kornfeld to lead, and there were lots of young people working around the clock, often showing up at three in the morning on a Saturday to spend six hours "bringing up" this assemblage of electronics to serve as a replica of our spacecraft flying to Mars so that we could spend the rest of the weekend testing it. (When you start your laptop, it takes about three seconds before you get that "bong" that tells you it's fully up and running. For us, that upload took a full shift of six or eight hours.)

This is where you come to appreciate that definition of genius (in this case, collective genius) as an infinite capacity for taking pains. Everything we were doing was hugely labor intensive, working it in finer and finer detail, double- and triple-checking in a regime that combined incredible tedium and incredible pressure. Anybody can come up with a wild idea. It's this 99 percent perspiration—implementation—that's in some sense the true test.

One of Miguel's great virtues is his capacity to hang at this level of detail and cut through the chaff to find the essential risks. In terms of navigation, EDL needed to perfect our coordination with Steve Lee's approach team. (Steve, who had managed the Guidance, Navigation, and Control effort that Miguel had been the chief engineer of, was a great friend and colleague, and a welcome choice to lead the effort for the approach to Mars.) How were we going to manage the handoff? How would any inaccuracies in navigation cause risks in EDL? How could we use our EDL capabilities to reduce some of the risks associated with errors in navigation on the way from Earth to Mars?

We had a series of dress rehearsals, each three or four days long and in ascending levels of "dress." The first was more like a table read when you're doing a play. It wasn't in real time, so we could stop and start and compress and expand. The second rehearsal was done in chunks that were time consistent. The third was in real time all the way, straight through, around the clock, with the 3 a.m. stuff happening at 3 a.m. We had weather reports to give us specific conditions and a gremlin team cooking up problems for us in simulations with complete data sets that we had to unravel. When we got to the real deal, we wanted no surprises.

For the last of these dress rehearsals, we went so far as to have everyone wear his or her official *Mars Science Laboratory* landing-night blue polo shirt so that we'd all look uniform on camera. Projections were that we'd have fifty million people tuning in. We felt that we had something important to communicate, and we wanted it to look as if we knew what we were doing.

Then, seventy-two hours before the landing, Miguel discovered the error in the center of navigation, which threw us into our last round of turmoil. By Saturday morning we had made our decision to implement an eleventh-hour change to the software parameters. The rest of the day went by as a stream of numbers flashing on computer screens in the mission control room.

That night, around ten o'clock, I went home to go to bed. I had been working on this project for nine years, in charge of EDL for seven. I loved the intensity of it, but the intensity meant that I had not gotten really good sleep in ages. That night I slept like a baby, and I don't really know why. I did not feel certain of success; I did not feel certain of failure. Perhaps it was the calm that comes over you when you know there is nothing more you can do. EDL was now in the hands of the fates, or God or whatever. We had done our best, and time would tell if our best was good enough.

Sunday morning dawned bright and warm. I remember making tea with my very pregnant wife and saying, "Okay, it's off to work. Let's see what happens."

EDL's day began with a 7:30 a.m. atmosphere meeting to go over the weather report from Mars. This was really just hand-wringing, because we would have to see a huge change in conditions to make any adjustments. Even so, we had half a dozen climate specialists going over weather data from *Mars Reconnaissance Orbiter,* making sure we knew what Mars was likely to throw at us during landing. The air on Mars is notoriously unstable. If you were standing on the surface, the temperature can vary by 40 degrees Fahrenheit between your feet and your head, and a dust devil can be born in an instant. Regional and even global dust storms can be spawned in a day or two. Luckily the forecast for Mars was fine and clear. Atmospherically speaking, you couldn't ask for a better day to land.

By afternoon there was not much going on at Mission Support but waiting, so I left the lab to get a moment away. I gave a talk at the Mars Society meeting in Pasadena, and then I went home to say hello to an aunt and uncle and cousins who were visiting.

Then it was back to JPL, where everyone was nervously munching peanuts, a JPL tradition since 1964, when peanut consumption supposedly helped a probe land on the moon after seven tries.

I gathered the team for a last locker-room pep talk. "The die is cast," I said, "and for better or for worse, we've done what we could. At least for me personally, when I look at what we've done, I think that we've done good. In fact, I think we've done great. So no matter what happens, be proud." We then went down to the JPL mall to take our picture beside the life-size rover model they had erected for the press.

Not everyone could break away, but all those who could followed me down en masse. The national media were all over our replica rover, and the network correspondents were doing their stand-ups in front of it, but we barged right in and essentially preempted their broadcasts to use the rover as the backdrop for our team picture. We hadn't even brought a photographer with us, because we knew there would be about a thousand cameras down there. As it came to pass, our good friend Tom Wynne from the JPL photo lab was on hand, so it was very much all in the family.

We posed—*snap!*—and then we all hugged one another. I told the team that whether we were headed for success or failure, it had been the greatest honor of my professional life to work with them. I meant it, every word. And then we all said, "See you on the other side." That's because we were going off in two different directions, some to what we call the war room and some to the cruise mission support area (MSA).

Then, just before eight, Miguel and I put on Sinatra.

The war room is where the bulk of the EDL team would be watching the data come rolling in, tracking in great detail how we were doing as we descended toward the Martian surface. The MSA is where the national cameras would be trained, where the theater would happen that would bestow upon us either glory or scorn, with mine as the one ass to kick.

It felt different as the rooms around the MSA continued to fill up with politicians, Caltech bigwigs, and senior managers from NASA and JPL. I had too much on my mind to pay much attention, but occasionally I would recognize an ex-governor or a senator.

It was now half past eight, and our screens were up with whatever view into the spacecraft data the individual wanted to look at, and in whichever format.

We had now crossed the threshold into lockdown, where we were not supposed to get up and move around. We were also within the time horizon for the national news, and at any moment we might be going to one of the networks with a live data feed. We didn't want a whole bunch of people picking their noses or shuffling around as if they didn't have anything to do, as if they weren't fully invested.

In two more hours, we'd be reaching the critical stage of EDL. So now it was "ears off," which meant that we commanded the spacecraft to stop listening to us. We didn't want any signal or noise—either accidental or maliciously intended—to do anything to take the capsule off course.

Nine minutes before we hit the atmosphere, we separated from the cruise stage. We de-spun the spacecraft from the long, twirling spiral that had kept it stable since shortly after launch and realigned so that the heat shield was in the right orientation to keep the whole module from becoming toast. The goal was to be pitched 15 degrees above horizontal, which is essentially the attitude of a jumbo jet coming in for a landing on a nice and easy glide path, nose up.

At 10:24 the aeroshell hit the atmosphere at around 13,000 miles per hour. Our heat shield was about to take the brunt of the friction, which would generate temperatures of up to 9,000 degrees Fahrenheit.

We were into it now, the terrifying moments when there's no data coming in and we're all alone with our thoughts, waiting for the next piece of news.

The first data we got said, "Beta out of bounds catastrophic," which was hardly reassuring. *Beta* refers to the yaw angle, or side-to-side orientation, for our angle of attack. *Catastrophic* meant more than 10 degrees off, suggesting that we were wobbling or careening and that the heat shield might not be pointed in the right direction, meaning that we might get the brunt of our heating on our back shell and burn up. But that first data burst came from the first few measurements, which occurred when the picture hadn't built up much detail, so the ratio of noise to signal was very high. Then the data resolved, and the message improved to the point that we could breathe again.

Happily there's now so much data coming at us that we don't have time to think about it. Then again, it's also way too late to do anything, because it would take a minimum of 13.8 minutes for news from Mars to reach us and for any responding signal from us to get back to the capsule.

So we're simply watching things happen, trying to ignore the fact that whatever we're seeing now happened almost 13.8 minutes ago. We're trying to live in the moment, observing and absorbing, saying "Okay, that's happened. . . . Okay, that looks great. . . . Okay, I'm waiting . . . ," and it's all coming at us in that same stream of blue numbers on dark computer screens, and I'm checking off the data points as we get lower and lower in the atmosphere, and I'm getting more and more nervous.

We're flying parallel to the surface and 7 miles up when our ridiculously huge parachute goes *kapoof,* bringing us down from 900 miles per hour to less than 200.

Then we jettison the parachute and turn on our rockets. We're over the landing site—or so the data stream from our radar tells us—and we begin our direct descent, rocketing downward at 100 miles an hour, then 50, then 15.

And now time is accelerating, and I feel as if I'm going down a ski trail, out of control, with everything whizzing by so fast that I can't keep up, and I can't do anything about it, so I'm just observing in a rush of sensory overload.

Our radar locks onto the target, and it likes what it sees. We're getting down to 200 meters (656 feet), the height of the tallest buildings in LA, and I keep waiting for the disaster to happen. And now each moment begins to dilate, and time slows to a crawl.

Cue Sky Crane. Time to head for the spotlight.

Al Chen, flight dynamics lead and the play-by-play voice over the PA, is reading telemetry off as he sees it, telling me and all the world at the same instant as we free fall, all or nothing at all.

"We're in Sky Crane," he announces.

It's happening. We're 22 meters (72 feet) from the surface, and we've slowed from 32 meters (105 feet) per second to a walking pace. Our rover the size of a subcompact has just dropped out of the descent stage and is rappelling downward as the bridles autowind, remaining taut as the larger descent stage continues its downward trajectory.

Then I hear the voice of God (it's actually Jody Davis, an EDL engineer from Langley, on the VOCA) saying, "Tango Delta nominal." *Tango Delta* is for "touchdown." *Nominal* is for "maybe." The computer for the descent stage, which is in the rover, has noticed that the throttle settings have dipped. The rover *thinks* it's landed.

Is the rover really sitting on Mars? The data feed tells us that it's begun to go through a set of housekeeping procedures, the most important of which is to cut the bridles. There's a watch spring at the top of the spool where the line comes in, and an anvil and a guillotine and a pyrotechnic charge. If all goes well, at the right moment—*kapow!*—the explosive charge drives that guillotine into the three nylon ropes simultaneously. The fourth cable is electrical, and it stays attached for a few more moments, then—*kapow!*—it gets cut and the vehicle flies away. Or so we hope. If the cuts aren't clean, we could drag the rover along on its side, or carry it away when the descent vehicle flies off.

Then I hear what sounds like "RIMU stable." *RIMU* is the acronym for "rover inertial measurement unit." *Stable* means that we haven't landed on a slope, and we're not moving sideways, as in dragging the rover. The descent stage has most likely detached.

I begin to count, "One . . . two . . . three . . . four . . ."

Brian Schratz, the radio guy, is counting off ten seconds on a stopwatch. Our last worry is that the descent stage, in a failure, will fall on the rover, but the first thing it would crush is the antenna. We still have our radio signal, which is UHF and coming from that antenna, so that's a good sign, but for it to mean anything, we need it to continue.

Have we done it? Has our insane contraption for landing in the Martian dust actually worked?

I'm counting along, but I'm stuck with four fingers in the air. I'm pacing. We're all supposed to sit, but fuck sitting. I'm pacing. I'm a pacer.

Brian gets to ten. He says, "UHF strong." If the antenna were going to be crushed, it would have been crushed by now.

Al's eyes are on me, waiting for the signal telling him he can say it.

Then it really settles over me. I can let go of all my dread and accept it. I give him the nod, and he says, "Touchdown confirmed. We are safe on Mars."

The room just goes crazy. Guys are hugging, and people are screaming, and I'm simply stunned. I give Mig this big hug, and all of a sudden I am just dazed, and I realize why tennis players fall down on the court when they win Wimbledon. This thing we've been working on as a team for ten freakin' years . . . and all of a sudden, it just happened! And it happened beautifully. It worked! It seems unreal, surreal—it just worked!

So what do I do now? I've been so focused for so long, and now I don't have to be focused anymore. Mig and I still have some housekeeping data to go over, but for a minute or two, we're just jumping up and down.

The biggest, and happiest, chore right now is to transfer control to the rover drivers over in the other building. From now on it's their baby. They will be living on Mars time, wearing 3-D glasses, and working while the rover sleeps to program every move *Curiosity* will make the next day in its exploration of the surface.

It feels like two seconds, but it's been forty minutes, and the landing team have already downed all their booze, and half of them have left to go to the press conference. But back at the launch, we'd taken two glass jars, one jar representing days to Mars and the other representing days since launch, and filled the days-to jar with marbles; every day, as part of the ritual of operating the spacecraft during its long cruise, the flight director would take a marble from the days-to jar and put it in the days-since jar. Now it was time to hand both jars over to the rover guys to start counting days on Mars. Keith Comeaux, the EDL flight director, and I did the honors.

I knew there was no way to re-create the storming of von Kármán from 2004. We shouldn't even try. That sort of thing works only when it's spontaneous. But I wanted my team to be at the press conference, and this time the bosses said, "Look, just come over with your team, but be quiet. Come over a little late, and we'll let everybody in."

But the guys were outside the doors once more and chanting "E D L!" and I was trying to shush them, saying, "Keep it down!" Then the director of communications sees me and says, "Adam, what the fuck are you doing? You're supposed to be up on the dais." This time out, they want me to speak.

So I tell my guys to chill for a while. I have been told that they will be let in later. I go in and mount the dais and listen to the speeches. Charlie Bolden, the head of NASA, talks, and then the president's science adviser talks. They are proud and eloquent, but to me it sounds like the same old blah-blah-blah.

So after a while I whip out my phone and I have a text from Al Chen outside, asking, "They won't let us in. Storm now?"

I don't write back for a moment, and I continue to listen to what seemed at the time to be some kind of disconnected political speech.

Al texts again, "Bring us in."

I respond, "Rush the door now."

Al texts back, "They won't let us. They're fighting us."

I text back, "Win."

Which Al, Mig, Steve, Devin, Dave, Jeremy, Jody, Paul, Fred, and the rest of the team do. And so, without planning to, we re-create the storming of von Kármán. They rush in, chanting, and it still feels slightly artificial, because it's a replay from 2004, but still we are incredibly excited, high-fiving across the dais. I wish Mig was up there with me.

Then, after the cheering dies to a point where we can get control of the room, the bunch of us sit down behind the dais. There are more speeches, but these feel more connected. Charles Elachi, JPL's director; then Pete Theisinger; and then Richard Cook. As Pete and then Richard speak, I look out over this auditorium, the faces of two hundred teammates, the press, the cameras, and I feel overwhelmed.

It had been a job, my job. A fucking awesome, full-throttle, use-my-whole-self job but a job nonetheless. I had not done this job with any conscious aspiration about furthering humanity or discovering the great unknown. I really had just wanted to make something happen, something cool. As I sit on the dais, I realize it was also something much, much more. It was an enormous human effort, it was something of our humanity with which I had been entrusted. Landing, our effort to explore, our struggle to understand our world, our universe, ourselves—it had all been rolled up in what we had just done.

Richard is speaking now, he is giving me credit, some of it due, but most of it feeling too much. I realize that I have not written anything to say at this moment. I never thought through what I might want to express. I have not even allowed myself to imagine

EDL really succeeding. Planning a celebratory speech would have felt like a dereliction of duty.

Richard pitches to me. I float for a moment in space, emotionally overloaded. "Say something profound," I hear him say.

I pull my emotions back, barely. I try to express how honored I am to have been part of this act of humanity, this effort that spanned more than one nation, across a decade, and involved the lives of thousands. How lucky I was to lead a team so capable and to live in a nation striving so gallantly to extend the edges of what humanity can do, of who we can be. I feel incredibly privileged and American.

I try to trot out my favorite riff on the Teddy Roosevelt quote, the one about working together being "the best prize life offers," but emotion robs me of my nouns, and I feel a little adrift. Finally I thank the "Blue Shirts," the team assembled, all in our landing-night finest. Certainly the finest group of people with whom I could ever hope to come together in working toward a common—but also very uncommon—goal.

EPILOGUE

FOR ALL THAT THE *CURIOSITY* LANDING MEANT TO ME AND our team, what I couldn't know at the time was how much it was going to mean to other people. The worldwide television coverage was a tip-off, as were the crowds filling Times Square at one in the morning on a Monday to watch our landing on the giant video screens.

But still I wasn't prepared for the onslaught of interview requests and invitations to speak to groups. Or the letters that started coming in from Egypt, China, India, all over the world. I got one from a science teacher in England who was a gardener and who'd read that I gardened. He asked for a memento, so I sent him one of the enameled pins I'd had made to give out to the team on landing night. Then there was the brassy lady who came up to me at the airport and asked me, "Are you Adam?" I said yes. She said, "You gave me the best night of my life!" Some tiny Central American country even created a postage stamp with a picture of Sky Crane on it.

My sense is that this enthusiasm isn't the same as when people get worked up over the World Cup or *Dancing with the Stars*. The people who contacted me described what they'd witnessed as being profound in ways that most media sensations are not. They seemed to get the idea that landing on Mars is the virtual extension of ourselves into space. To use a term I learned from Jebediah Springfield of *The Simpsons*, I think they sensed that exploring beyond our home planet "embiggens" all of us.

I know for sure that it embiggened me.

Three weeks after the landing, my daughter Olive was born, which seemed to complete the transformation of my personal life, which in turn rounded out the transformation of my professional life. The crazed and compartmentalized guy of MER was no more. I'd experienced a divorce and a remarriage, and through a lot of conscious effort, I think I'd gained a much more unified psyche. The success of MSL put me in a position of emotional calm that left me free and able to try to play any role in the creative process without worry. I could stay in the background if that was what was called for. I could be the hero or the villain—it didn't matter.

In 2013 I was made a Fellow at JPL, which meant that my only career concern going forward would be the long-term strength and success of the lab, as well as our nation's exploration of space. This means helping to keep our programs on the most productive path technically but also mentoring younger engineers and speaking up for the people I think are really good.

All the same, my primary technical niche remains EDL. I'm working on a new supersonic parachute technology that can help land humans on Mars, as well as novel robotics approaches to dealing with the icy and uncertain terrain found on the surfaces of Europa and Enceladus (moons of Jupiter and Saturn, respectively), high-priority targets for our continued exploration. I'm also working on a mission to go back to Mars in 2020 using the same *Curiosity* landing system but for a different purpose. *Curiosity* was sent to scout out and analyze the surface. The plan is for the new rover we're developing to be part of that long-awaited sample-return mission to Mars, selecting and acquiring samples from the surface to be brought back to Earth. Ultimately we're going to put humans on Mars, which is why I'm also helping to develop the infrastructure they'll need, which includes rocket-fuel-generation plants and nuclear power plants.

Miguel is working on a sample-and-return mission to the comets. Tom Rivellini is now at Apple, inventing cool stuff that, sad to say, he can't tell me about. Allen Chen is leading the EDL effort for the Mars 2020 mission, and Richard Cook is managing the Solar System Exploration Directorate, spreading his talents across the other non-Mars solar system destinations. Dara Sabahi is now chief engineer of the Engineering and Science Directorate, and although our days of working together may have passed, I will forever be indebted to the lessons he taught me. We are all moving forward. Nobody is slowing down. Nobody seems complacent. The passion and the curiosity keep on going for all of us.

At this point in my career, I do spend more time than I used to thinking about what exactly it is that we do when we explore and why we do it in the first place. Enthusiasts for the space program have always touted the indirect, practical benefits, like the way JPL's worldwide radio tracking network helped drive the emergence of the telecommunications industry. This trickle-down technological benefit exists, but in my mind, any immediate, practical payoff is way down the list of reasons why we continue to explore space.

You can look at commercial air travel, for instance, as an obvious, practical outcome of our long exploration of flight. When I speak to groups, I often show on the screen behind the podium a selfie of me boarding the plane to come to them. But then I follow up with some old footage taken in the late nineteenth century, when finding a way to get an engine and wings to work together was still the cutting edge. You see men in bow ties and bowler hats jumping off bridges with kites strapped to their backs, and suicide machines that launch about 3 feet off the roof of a barn before nosediving into the ground.

If that early experimentation was driven by practical considerations, it certainly didn't look it at the time. It turned out that there were fantastic applications waiting down the road, but for the individuals involved in the birth of aviation, what they were doing was clearly at odds with the most practical consideration a person can have, which is living long enough to make it home for dinner. These would-be pilots were being kind of crazy, but in a way that I see as an inescapable expression of their humanity. They were driven to explore flight, not because it was practical but despite its being impractical. They were driven to "find out," to know more—searching to become more knowledgeable, more powerful, simply *more*.

When the first engineer came up with the first wheel, or the first throwing stick to improve spear power, I think the impulse was very similar. That drive led the first artist to make the first cave paintings. It has given us art and literature, philosophy and religion, the jet airliner, fracking, penicillin, and the smartphone. I feel lucky to be an engineer, a most human profession—one part creator, one part understander. As Theodore von Kármán, the first director of the lab, put it, "Scientists study the world that exists; engineers create the world that never was."

We humans are an innately curious species. Born through hips too narrow to pass a skull large enough to hold a fully formed human brain, we are born half-baked. Compared with those of other animals, very few of our behaviors are hardwired. We don't inherit genetic instructions for nest building, for instance, or for migrating south when the sun hits a certain angle in the sky. We come into this world programmed with few instructions, save for one paramount piece of code: Be curious.

Following that one command, each of us begins to construct her understanding of the universe. I see it in my two-year-old daughter, and I see it in my seventy-year-old colleague. We all share the instinct, and we can all relate to it, which gives our curiosity-driven need to explore the added beauty that no one has to be forced to do it. People don't necessarily have to be incentivized with stock options and mega-salaries for curiosity-driven creativity to occur. The only essential ingredient is a work environment that's structured to encourage our innate drive to wonder, question, and explore. Just as each of us built our understanding of our world during childhood, given the chance, everyone will follow the lure of their own curiosity and their own desire for mastery throughout their working lives.

Work that contains that kind of inherent incentive is one of the most powerful drivers we have, for innovation and decision making founded on curiosity outperform the same based on fear. Evolutionary biologists talk about exploiting and exploring, the two complementary survival strategies that every anthill or beehive seems to understand. Every organism, or colony, or business, for that matter, needs to find the right balance between making the most of existing resources and the investment necessary to keep scouts out exploring the edge.

We explore as a gesture of our humanity. We do it because we can, and we do it as an affirmation of who and what we are. As a society, if we ever stop exploring, who will we be? I think we will be stagnant, not innovating, not building. It's a formula not just for stagnation but for disaster. Which is why nurturing and supporting innate curiosity is still one of the most valuable survival tools we have.

Over the past seven million years, our species has expanded our range to include the whole planet, and much of that migration must have looked crazy at times. In the past fifty years, we've landed on the moon and reached out to Mars. And all along the way, we've been spreading our genes as well as our memes—the ideas and techniques and traditions that are the cultural equivalent of DNA.

We carried out this exploration and expansion in teams and in bands, but we didn't do it in lockstep. Loyalty and cooperation were essential, but so were individualism and creative conflict. But keeping the conflict creative, rather than combustible, was the search for the truth that lets ideas win, not people. When we were most successful, it was because we were trying to find what was right, not to be right. Our curious search for truth and understanding makes us unique compared with the other creatures on this planet. Our search drives strange behaviors, creates strange inventions, and appears crazy at times, but it just might be the right kind of crazy.

ACKNOWLEDGMENTS

We are social beings and who we are is part genes and part some really fantastic accumulation of the influences of those in our lives. Much of the story contained within these pages takes place at Caltech's Jet Propulsion Laboratory. JPL is a collection of the finest people I have ever known. Of the many strokes of luck that I have experienced in my life, perhaps one of the most profound was to find myself a member of the technical staff at JPL. The lab has been a great teacher, a part-time therapist, and a laboratory in which I was safe to perform experiments on who I would become. I have been substantially shaped by my time there and am eternally grateful to the many wise advisers, mentors, and friends I have found within its gates.

This book would not have been possible without the help of my loving and understanding wife, Trisha; and my daughters, Caledonia and Olive. Their patience allowed this work to take form. The book stands as it does through the tremendous efforts of my coauthor, Bill Patrick, who built the bones of this story; and from the countless, but brilliantly posed, questions of my editor, Niki Papadopoulos, which drew out the best in my thinking. This whole thing began with the first call to my now favorite literary agent, Daniel Greenberg, some three years ago. Thank you, Daniel, for setting up and smoothing out a great adventure. Along the way I have searched out advice from friends and colleagues who have read and helped me keep a balanced eye. I thank them all but none of them has helped as much as my friend Mark Davis—thank you, Mark. It has been said that it takes a village to raise a child, or perhaps build a spacecraft, or write a book. If this is so, I am a lucky man and quite happy in my village.

INDEX